Chemistry of Asphaltenes

Chemistry of Asphaltenes

James W. Bunger, EDITOR

University of Utah

Norman C. Li, EDITOR

Duquesne University

Based on a symposium sponsored by the Division of Petroleum Chemistry at the 178th Meeting of the American Chemical Society, Washington, D.C., September 10–11, 1979.

ADVANCES IN CHEMISTRY SERIES 195

AMERICAN CHEMICAL SOCIETY
WASHINGTON, D.C. 1981

Library of Congress CIP Data

Chemistry of asphaltenes.

(Advances in chemistry series, ISSN 0065-2393; 195)

Includes bibliographies and index.

1. Asphaltene—Congresses.

I. Bunger, James W., 1945– . II. Li, Norman C., 1913– . III. American Chemical Society. Division of Petroleum Chemistry. IV. Series.

QD1.A355 no. 195 [TP692.4.A8] 540s [665'.4]
ISBN 0-8412-0592-2 81-19053 AACR2
ADCSAJ 195 1-260 1981

PRINTED IN THE UNITED STATES OF AMERICA

Advances in Chemistry Series

M. Joan Comstock, *Series Editor*

FOREWORD

ADVANCES IN CHEMISTRY SERIES was founded in 1949 by the American Chemical Society as an outlet for symposia and collections of data in special areas of topical interest that could not be accommodated in the Society's journals. It provides a medium for symposia that would otherwise be fragmented, their papers distributed among several journals or not published at all. Papers are reviewed critically according to ACS editorial standards and receive the careful attention and processing characteristic of ACS publications. Volumes in the ADVANCES IN CHEMISTRY SERIES maintain the integrity of the symposia on which they are based; however, verbatim reproductions of previously published papers are not accepted. Papers may include reports of research as well as reviews since symposia may embrace both types of presentation.

CONTENTS

1

On the Molecular Nature of Petroleum Asphaltenes

JAMES G. SPEIGHT[1]

Corporate Research–Science Laboratories, Exxon Research and Engineering Company, P.O. Box 45, Linden, NJ 07036

SPEROS E. MOSCHOPEDIS

Coal Research Department, Alberta Research Council, 11315–87th Avenue, Edmonton, Alberta T6G 2C2, Canada

A survey of the methods used to determine asphaltene "structure" indicates that there are serious shortcomings in all of the methods because of the assumptions required to derive the molecular formulae. The continued insistence that a complex fraction such as asphaltenes, derived in a one-step process from petroleum as a consequence of its insolubility in nonpolar solvents, has a definitive molecular structure is of questionable value to petroleum technology, and it is certainly beyond the scope of the available methods to derive such formulae. Asphaltenes would best be described in terms of several structural types rather than definite molecular structures.

The molecular nature of the asphaltene fractions of petroleum and bitumens has been the subject of numerous investigations (*1, 2*) but determining the actual structures of the constituents of the asphaltene fraction has proved to be difficult. It is, no doubt, the great complexity of the asphaltene fraction that has hindered the formulation of the individual molecular structures. Nevertheless, the various investigations have brought to light some significant facts about asphaltene structure; there are indications that asphaltenes consist of condensed aromatic nuclei that carry alkyl and alicyclic systems with heteroelements (i.e., nitrogen, oxygen, and sulfur) scattered throughout in various, including heterocyclic, locations.

Other basic generalizations also have been noted; with increasing molecular weight of the asphaltene fraction, both aromaticity and the proportion of heteroelements increase (*3, 4, 5*). In addition, the proportion of asphaltenes in

[1]Author to whom correspondence should be sent.

0065-2393/81/0195-0001$05.00/0

petroleum varies with source, depth of burial, the specific (or API) gravity of the crude oil, and the sulfur content of the crude oil as well as a nonasphaltene sulfur (*5*). However, many facets of asphaltene structure still remain unknown, and it is the purpose of this chapter to bring together the pertinent information on asphaltene structure as well as the part played by asphaltenes in the physical structure of petroleum and bitumens.

Asphaltene Composition

Asphaltenes are dark brown to black friable solids that have no definite melting point, and when heated, usually intumesce, then decompose leaving a carbonaceous residue. They are obtained from petroleums and bitumens by addition of a nonpolar solvent (such as a hydrocarbon) with a surface tension lower than 25 dynes cm^{-1} at 25°C (such as liquefied petroleum gases, the low-boiling petroleum naphthas, petroleum ether, pentane, isopentane, and hexane) but are soluble in liquids having a surface tension above 25 dynes cm^{-1} (such as pyridine, carbon disulfide, carbon tetrachloride, and benzene) (*6, 7*).

Asphaltenes are, by definition, a solubility class (*8, 9, 10*) that is precipitated from petroleums and bitumens by the addition of a minimum of forty volumes of the liquid hydrocarbon. In spite of this, there are still reports of "asphaltenes" being isolated from crude oil by much lower proportions of the precipitating medium (*11*), which leads to errors not only in the determination of the amount of asphaltenes in the crude oil but also in the determination of the compound type. For example, when insufficient proportions of the precipitating medium are used, resins (a fraction isolated at a later stage of the separation procedure by adsorbtion chromatography) also may appear within the asphaltene fraction by adsorbtion onto the asphaltenes from the supernatant liquid and can be released by reprecipitation in the correct manner (*12*). Thus, questionable isolation techniques throw serious doubt on any conclusions drawn from subsequent work done on the isolated material.

The elemental compositions of asphaltenes isolated by use of excess (greater than 40) volumes of *n*-pentane as the precipitating medium shows that the amounts of carbon and hydrogen usually vary over only a narrow range: 82 ± 3% carbon; 8.1 ± 0.7% hydrogen (*13, 14*). These values correspond to H/C ratios of 1.15 ± 0.05, although values outside of this range sometimes are found. The near constancy of the H/C ratio is surprising when the number of possible molecular permutations involving the heteroelements are considered. In fact, this property, more than any other, is the cause for the general belief that unaltered asphaltenes from virgin petroleum have a definite composition and that asphaltenes are precipitated by hydrocarbon solvents because of this composition, not only because of solubility properties. However, notable variations do occur in the proportions of the heteroelements, in particular in the proportions of oxygen and sulfur. Oxygen contents vary from 0.3% to 4.9%, so the O/C ratios vary from 0.003 to 0.045; sulfur contents vary from 0.3% to 10.3%, so S/C ratios range from 0.001 to 0.049.

Table I. Elemental Compositions of Asphaltene Fractions Precipitated by Different Solvents

Source	Precipitating Medium	Composition (wt %)					Atomic Ratios			
		C	H	N	O	S	H/C	N/C	O/C	S/C
Canada	*n*-pentane	79.5	8.0	1.2	3.8	7.5	1.21	0.013	0.036	0.035
	n-heptane	78.4	7.6	1.4	4.6	8.0	1.16	0.015	0.044	0.038
Iran	*n*-pentane	83.8	7.5	1.4	2.3	5.0	1.07	0.014	0.021	0.022
	n-heptane	84.2	7.0	1.6	1.4	5.8	1.00	0.016	0.012	0.026
Iraq	*n*-pentane	81.7	7.9	0.8	1.1	8.5	1.16	0.008	0.010	0.039
	n-heptane	80.7	7.1	0.9	1.5	9.8	1.06	0.010	0.014	0.046
Kuwait	*n*-pentane	82.4	7.9	0.9	1.4	7.4	1.14	0.009	0.014	0.034
	n-heptane	82.0	7.3	1.0	1.9	7.8	1.07	0.010	0.017	0.036

In contrast, the nitrogen content of the asphaltenes appears to remain relatively constant; the amount present varies from 0.6% to 3.3% at the extremes, and N/C ratios are usually about 0.015 ± 0.008. However, exposing asphaltenes to atmospheric oxygen can substantially alter the oxygen content, and exposing a crude oil to elemental sulfur, or even to sulfur-containing minerals, can result in excessive sulfur uptake. Perhaps oxygen and sulfur contents vary more markedly than do nitrogen contents because of these conditions.

In addition, the use of heptane as the precipitating medium yields a product that is substantially different from the pentane-insoluble material (Table I). For example, the H/C ratios of the heptane precipitate are markedly lower than those of the pentane precipitate, indicating a higher degree of aromaticity in the heptane precipitate. N/C, O/C, and S/C ratios are usually higher in the heptane precipitate, indicating higher proportions of the heteroelements in this material (*13*, *14*).

Asphaltene Structure

Asphaltene Structure by Spectroscopic Methods. Much of the information available on the carbon skeleton of asphaltenes has been derived from spectroscopic studies of asphaltenes isolated from various petroleums and natural asphalts (*1*, *2*). The data from these studies support the hypothesis that asphaltenes, viewed structurally, are condensed polynuclear aromatic ring systems bearing alkyl sidechains. The number of rings apparently varies from as low as six in smaller systems to fifteen to twenty in more massive systems (*13*, *14*).

Attempts have also been made to describe the total structures of asphaltenes (Figure 1) in accordance with NMR data and results of spectroscopic and analytical techniques, and it is difficult to visualize these postulated structures as part of the asphaltene molecule. The fact is that all methods employed for structural analysis involve, at some stage or another, assumptions that, although based on data concerning the more volatile fractions of petroleum, are of questionable validity when applied to asphaltenes.

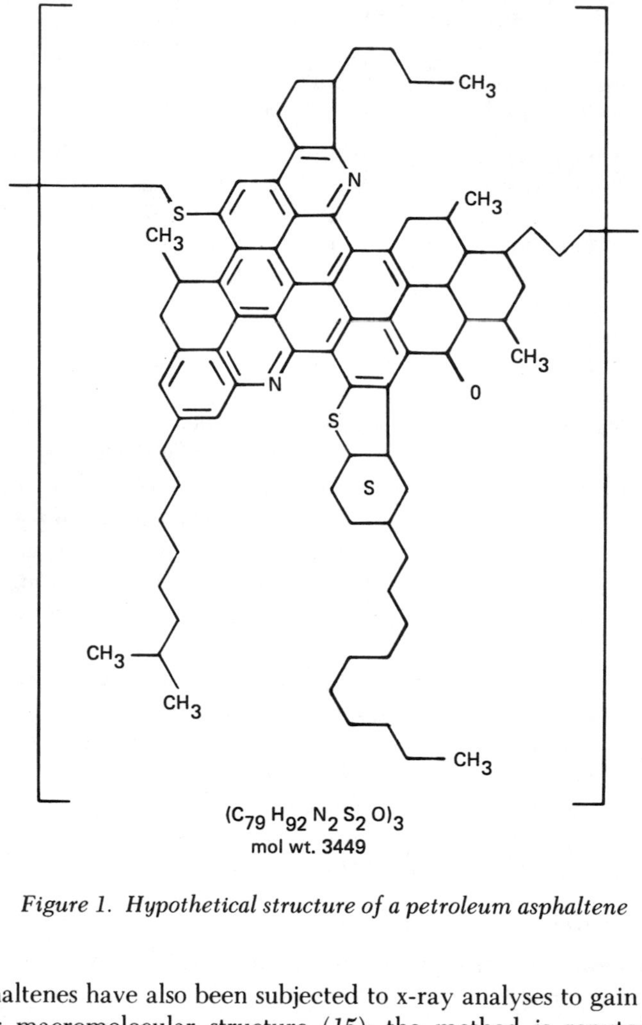
CH3
N
CH3
S
CH3
N
O
CH3
S
S
CH3
CH3
CH3
$(C_{79}H_{92}N_2S_2O)_3$
mol wt. 3449

Figure 1. Hypothetical structure of a petroleum asphaltene

Asphaltenes have also been subjected to x-ray analyses to gain an insight into their macromolecular structure (*15*); the method is reputed to yield information about the dimension of the unit cell such as interlamellar distance ($c/2$), layer diameter (L_a), height of the unit cell (L_c), and number of lamellae (N_c) contributing to the micelle (Figure 2).

Fractionation of an asphaltene by stepwise precipitation with hydrocarbon solvents (heptane to decane) allows separation of the asphaltene by molecular weight. The structural parameters determined using the x-ray method (Table II) show a relationship to the molecular weight (*16*). For the particular asphaltene in question (Athabasca), the layer diameters (L_a) increase with molecular weight to a limiting value; similar relationships also appear to exist for the interlamellar distance ($c/2$), micelle height (L_c), and even the number of lamellae (N_c) in the micelle.

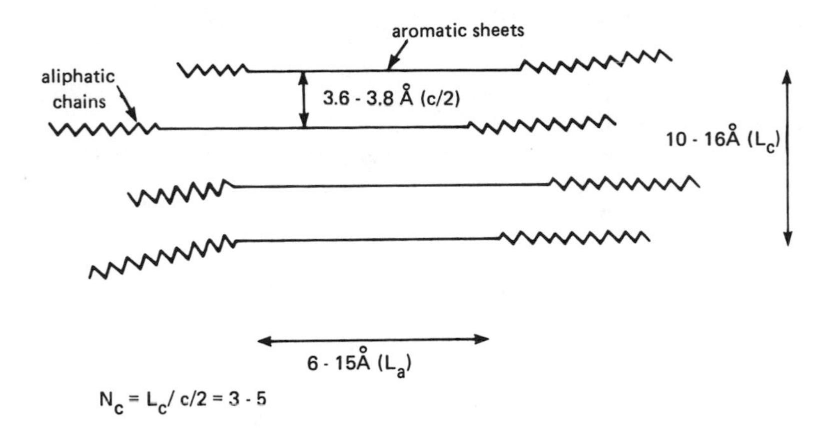

Analytical Chemistry

Figure 2. Representation of an asphaltene from x-ray analysis

Application of Diamonds's x-ray diffraction matrix method to the problem of determining asphaltene structure produced some novel results (*17*). For example, asphaltenes (precipitated by *n*-pentane from Athabasca bitumen and conventional Alberta crude oils) that are soluble in decane gave histograms completely different from those obtained with the decane-insoluble material (Figure 3), suggesting the existence of at least two different molecular types in the asphaltene fraction. The molecular types predominant in the decane-soluble material appear to be based on simple condensed aromatic units with only about six aromatic rings per unit. On the other hand, the decane-insoluble fraction contained aromatic systems of much greater complexity, but these systems appeared to be collections of simpler units that were actually similar to those in the lower molecular weight material but linked head-to-tail in a cyclic system.

Table II. Structural Parameters of Asphaltene Fractions Derived by the X-ray Method

	Structural Parameters			
Molecular Weight	L_a(Å)	*c/2* (Å)	L_c(Å)	N_c
2694	6.0	3.81	10.7	2.8
2704	6.3	3.79	10.2	2.7
3185	7.3	3.79	11.0	2.9
4338	9.1	3.74	11.6	3.1
6427	13.9	3.65	12.5	3.4
6530	14.5	3.65	13.8	3.8
7603	14.7	3.64	13.8	3.8
8158	14.8	3.64	14.0	3.9

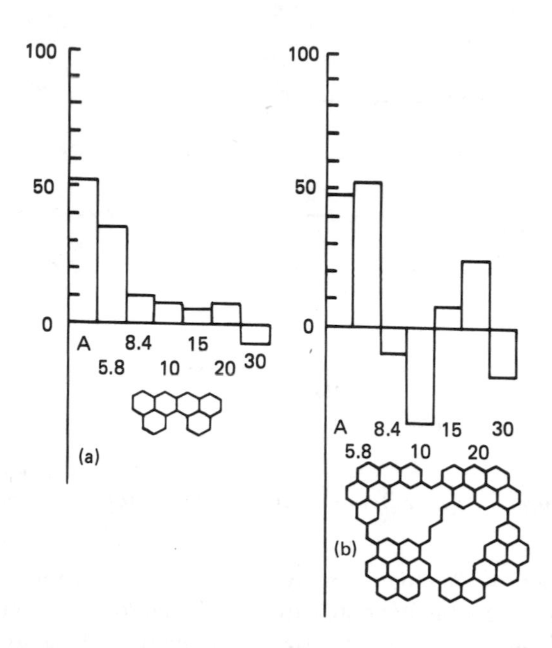

Figure 3. Histograms and hypothetical aromatic units for (a) decane-soluble asphaltenes, and (b) decane-insoluble asphaltenes from Alberta crude oils

In the case of asphaltenes of lower molecular weight, the absence of any evidence in the histograms to support hole structures clearly suggests that any heteroatoms located in the cyclic systems are found at peripheral sites. However, there is some evidence of hole structures within the carbon lamellae of the higher molecular weight asphaltenes. To accommodate heteroatoms in such holes would require a significant modification of the NMR model to a much larger, more ponderous structure. In contrast, insertion of the heteroelements into the x-ray model (Figure 3) could be achieved quite conveniently without any significant structural alterations. However, structural acceptability is not the only criterion for proof of structure when deducing structural formulae by any of the physico-chemical techniques. Indeed, the method itself may require proof of acceptability.

Investigations of the x-ray diffraction patterns of various low temperature (450°–750°C) synthetic carbons (*18*), carbon black blended with polyethylene (*15*), condensed aromatics of known structure where the maximum diameter of the sheets is approximately 14 Å (*15*) as well as mixtures of condensed aromatics and porphyrins (*19*) indicate that the x-ray diffraction patterns can be reproduced thereby supporting the concept of condensed aromatic sheets (having a tendency to stack) as the structure of asphaltenes. However, it is perhaps this ease with which the x-ray diffraction of the asphaltenes can be reproduced which dictates that caution is necessary in the interpretation of the data. Indeed, any empty polyethylene sample holder will exhibit a similar x-ray diffraction pattern to that of the asphaltenes (*19*). Therefore, it appears debatable whether the data derived by this method, or by any other method, are absolute or their sole value is limited to comparison between the various parameters.

Indeed, an excellent example of inconsistency between a spectroscopic method (i.e., NMR) and other data is provided by an examination of the asphaltenes from Athabasca bitumen where alkyl sidechains are deduced to contain approximately four carbon atoms (*20, 21*). The pyrolysis (350°–800°C) of this asphaltene produces substantial amounts of alkanes ($\leq C_{34}$) in the distillate (*22, 23, 24*). The presence of these alkanes in the pyrolysates is thought to reflect the presence of such chains in the original asphaltene (*24*) but this is difficult to rationalize on the basis of an "average" structure derived from NMR data. Obviously, recognition of the inconsistencies of the spectroscopic method with respect to the paraffinic moieties must lead to recognition of similar inconsistencies when considering the aromatic nucleus.

A key feature in the current concept of asphaltene structure is believed to be the occurrence of condensed polynuclear aromatic clusters, which may contain as many as twenty individual rings and account for approximately 50% of the asphaltene carbon (*1, 2, 13, 14*). However, it would be naive to presume that precise (or meaningful "average") molecular structures can be deduced by means of any spectroscopic technique (*1, 2, 13, 14*) when too many assumptions (incorporating several unknown factors) are required to derive the structural formulae.

Asphaltene Structure by Chemical Methods. The concept of asphaltenes being a sulfur polymer of the type:

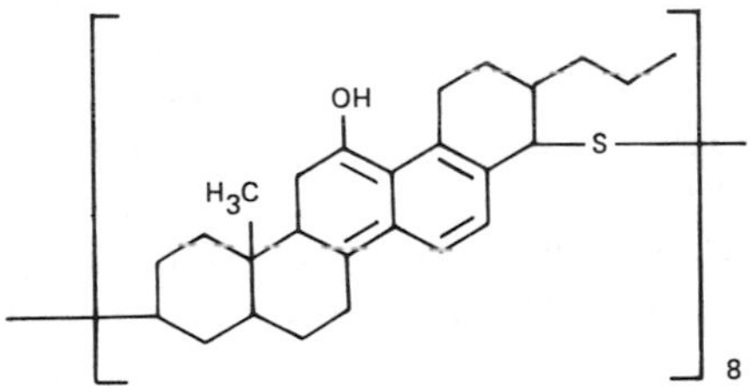

OH
H_3C
S
8

or even a regular hydrocarbon polymer of the type:

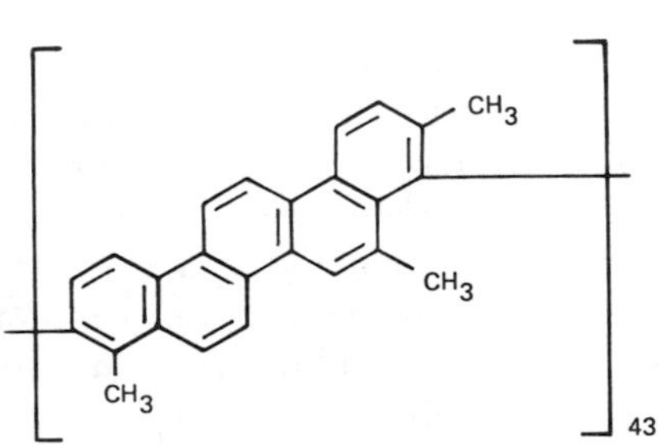

CH_3
CH_3
CH_3
43

has arisen because of the nature of the products obtained by reacting an asphaltene with potassium naphthalide (*25*). However, it was erroneously assumed that this particular organometallic reagent—one of several known to participate in rapid, complex reactions with organic substrates—cleaved only carbon–sulfur–carbon bonds, and not carbon–carbon bonds, that is:

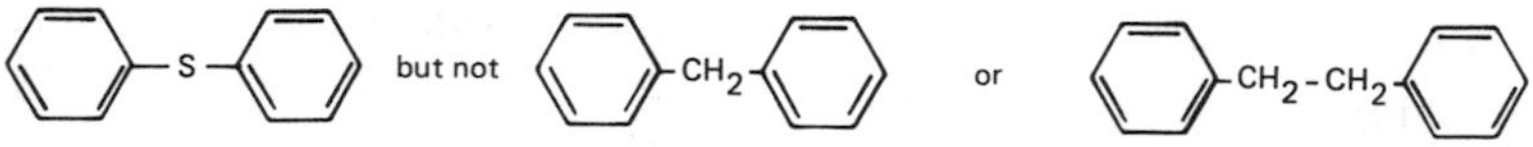

S but not CH_2 or CH_2-CH_2

It has been proved that potassium naphthalide cleaves carbon–carbon bonds in various diphenylmethanes (*26*) and the cleavage of carbon–carbon bonds in 1,2-diarylethanes has been documented (*27*). In each case the isolation of well-defined organic reaction products confirms the nature of the reaction. Furthermore, the possibility of transmetallation from the arylnaphthalide to the aromatic centers of the asphaltenes (*27*) complicates the situation and undoubtedly leads to more complex reactions and to reaction products of questionable composition. Formulating the structure of the unknown reactant (asphaltenes) under such conditions is extremely difficult. Indeed, it is evident that reacting asphaltenes with any particularly active reagents (e.g., the alkali aryls) leads to complex reactions, and it may be difficult, if not impossible, to predict accurately the course of these reactions (*28*). In fact, the reaction of potassium naphthalide with tetrahydrofuran alone—under conditions identical to those reported where asphaltenes were also present (*25*)—produces a light brown amorphous powder (*29*) that could erroneously be identified as a major product had any asphaltene been present.

There have also been attempts to substantiate the concept of a sulfur-containing polymer by virtue of the thermal decomposition in the presence of tetralin (*30*). However, aliphatic carbon–carbon bonds (such as those in 1,2-diphenylethane) will cleave under similar conditions, while reactions involving alteration of the hydrocarbon structure also occur (*31*). Indeed, an investigation of the nature of the thermal dissociation of tetralin indicates that the reaction is quite complex (*32*) and the presence of an added material (such as coal or asphaltene) could render any attempt to rationalize the reaction in simple terms to be of extremely dubious value.

Nitrogen, Oxygen, and Sulfur in Asphaltenes. Unfortunately, in all these studies, too little emphasis has been placed on determining the nature and location of the nitrogen, oxygen, and sulfur atoms in the asphaltene structure. However, mass spectroscopic investigations (*33*) of a petroleum asphaltene have allowed the identification of fragment peaks, which indicate that at least some of the heteroatoms exist in the ring systems (Figure 4). A study of the thermal decomposition of the asphaltenes from a natural bitumen (*22*, *23*) indicated that only 1% of the nitrogen was lost during the thermal treatment, while substantially more sulfur (23%) and almost all of the oxygen (81%) were lost as a result of this treatment. The tendency for nitrogen and

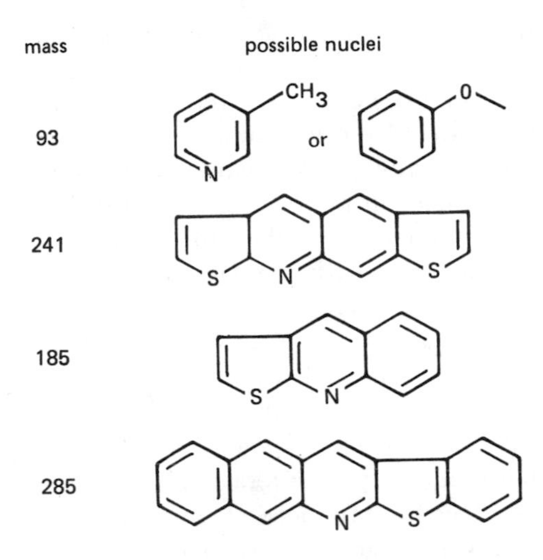

Analytical Chemistry

Figure 4. Suggested heteroatom locations from mass spectroscopic fragmentation peaks

sulfur to remain during thermal decomposition, as opposed to the easy elimination of oxygen, supports the concept that nitrogen and sulfur have stability because of their location in ring systems (*24, 34, 35*); if a sulfur-type polymer (*25*) existed, it would be expected to lose much, but not necessarily all, of its sulfur content to the volatiles—mainly as hydrogen sulfide (*24*).

Such a sulfur-type polymer might also be expected to lose much of the sulfur by treatment with Raney nickel (*36*). However, these particular (Athabasca) asphaltenes are difficult to desulfurize with Raney nickel (*26*) as compared with a variety of aromatic/aliphatic thioether polymers of the type:

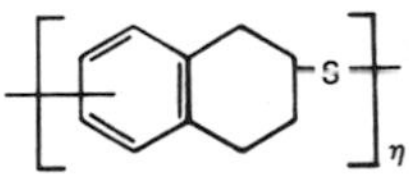

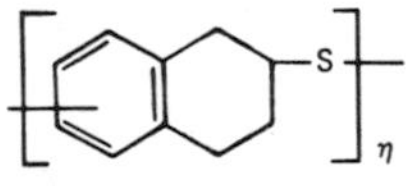

which can lose sulfur quite conveniently under the same conditions (*37, 38*). Polymers of the type:

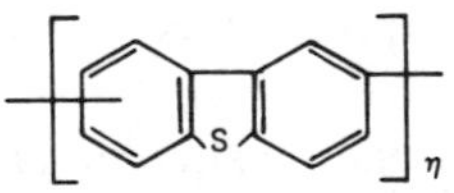

desulfurize very slowly with Raney nickel (*37, 38*) and others, for instance:

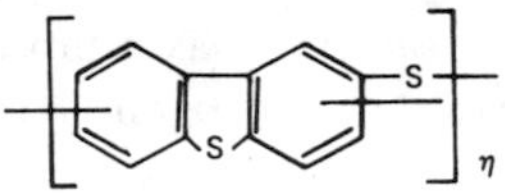

lose nonring sulfur relatively readily compared with the heterocyclic sulfur (*37, 38, 39*).

If these results, with the data from thermal decomposition experiments (*22, 23, 24*), can be projected to the particular asphaltenes in question, it would be indicative of sulfur existing predominantly in the heterocyclic form.

Of the heteroelements in petroleum, there are more data pertaining to the locations of the oxygen atoms than to the sulfur and nitrogen atoms. However, the majority of the data relates to oxygen functions in blown (oxidized) asphalts and residua (*40, 41*), which may be of little relevance to the oxygen functions in the native materials. Of the limited data available, there are indications that oxygen in asphaltenes can exist as nonhydrogen-bonded phenolic hydroxyl groups (*42, 43*). Other information on the presence and nature of oxygen in asphaltenes has been derived from infrared spectroscopic examination of the products after interaction of the asphaltenes with acetic anhydride (*42, 44*). This has produced evidence for the presence of ketones and/or quinones as well as indications that a considerable portion of the hydroxyl groups present in asphaltenes may exist as collections of two or more hydroxyl functions on the same aromatic ring, or on adjacent peripheral sites on a condensed ring system:

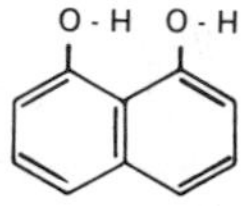

or even on sites adjacent to a carbonyl function in a condensed ring system:

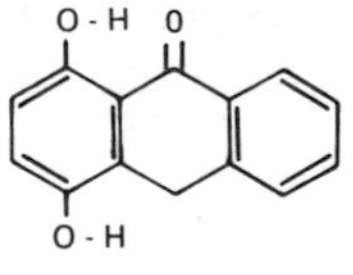

In the context of polyhydroxy aromatic nuclei existing in Athabasca asphaltenes, it is of interest to note that pyrolysis at 800°C results in the formation of resorcinols (*24*) implying that such functions may indeed exist in the asphaltenes.

The form in which nitrogen exists in asphaltenes is even less well understood than those of oxygen and sulfur. Attempts have been made to define organic nitrogen in terms of basic and nonbasic types (*45*). Spectroscopic investigations (*46*) suggest that carbazoles might be a predominant nitrogen type in asphaltenes from Athabasca bitumen, which supports earlier mass spectroscopic evidence (*33*) for nitrogen-types in asphaltenes.

Asphaltene Molecular Weights

Determining the molecular weights of asphaltenes is a problem because they have a low solubility in the liquids often used for determination. Also, adsorbed resins lead to discrepancies in molecular-weight determination, and precipitated asphaltenes should be reprecipitated several times prior to the determination (*12*). Thus, careful precipitation and careful choice of the determination method are both very important for obtaining meaningful results.

Asphaltenes tend to associate, even in dilute solution (*47, 48*) so there has been considerable conjecture about the actual molecular weights of these materials, but the data are fragmented and many different asphaltenes have been employed for the investigations. Nevertheless, for any one particular study, there are large variations in asphaltene molecular weights. For example, molecular weight investigations using an ultracentrifuge gave values up to 300,000 (*47, 49, 50*) while an osmotic pressure method (*51*) indicated molecular weights of approximately 80,000 and a monomolecular film method (*52*) yielded values of 80,000–140,000. However, other procedures have yielded lower values: 2500–4000 by the ebullioscopic method (*53*); 600–6000 by the cryoscopic method (*51, 54–57*); 900–2000 by viscosity determinations (*49, 58, 59*); 1000–4000 by light adsorption coefficients (*60*); 1000–5000 by vapor pressure osmometry (*5, 61*); and 2000–3000 by an isotonic (*62*) or equal vapor pressure (*63*) method.

A fairly comprehensive study of asphaltene molecular weights by vapor pressure osmometry has been reported and shows that the molecular weights of various asphaltenes are dependent not only on the nature of the solvent but also on the solution temperature at which the determinations were performed (*64*). However, data from later work, involving molecular weight determinations by the cryoscopic method (*48*), indicate that the molecular nature of asphaltenes is not conducive to the determination of absolute molecular weights by any one method. For any one method, the observed molecular weights suggest that asphaltenes form molecular aggregates, even in dilute solution, and this association is influenced by solvent polarity, asphaltene concentration, and the temperature of the determination. In fact, work by Ignasiak et al. (*65*), has confirmed the earlier work of Moschopedis and Speight (*66*) and Moschopedis et al. (*64*), which showed that intermolecular hydrogen-bonding is involved in asphaltene association and has a significant effect on observed molecular weights. This requires that serious reconsideration be given to ideas such as the concept of these particular asphaltenes existing as a sulfur polymer (*25*). This sulfur-polymer concept, in spite of evidence to the contrary (*64*), was based on the erroneous statement that solvent type did not affect asphaltene molecular weight and could not account for the observed decreases in molecular weight with increases in solvent polarity.

The Physical Structure of Asphaltenes in Petroleum

Although the evidence available in the literature appears to indicate that the hydrocarbon structures and some features, such as the various condensed ring systems, in different petroleums are similar (from the asphaltenes and resins to the constituents of the oil fraction), the variety of source materials involved in petroleum genesis implies that, on an individual molecular scale, there may be substantial structural differences among the constituents of the various crude oils and bitumens. As well, the difficulty with which resins from one crude oil peptize (As in a colloid, the terms peptization, dispersion, and solubilization are often used interchangeably to describe the means by which asphaltenes exist within petroleum.) asphaltenes from a different crude oil, and the instability of the "blend" (*5*) are evidence for significant structural differences among the asphaltenes and resins of various crude oils.

Some researchers, in addition to deriving broad generalities from the use of various analytical techniques, are attempting to assign specific total molecular configurations to the asphaltene constituents. Perhaps it is of little value to petroleum technology and certainly beyond the scope of the available methods to derive such formulae. However, within any one petroleum, a variety of structures exist in the asphaltene fraction (in which there is a decided hydrogen deficiency), but the close relationships of the various hydrocarbon series comprising the asphaltenes, resins, and oils gives rise to much overlapping of fractions into neighboring series, both in molecular weight and in H/C ratio.

The asphaltenes appear to be the final (excluding carbenes and carboids, those organic fractions of petroleum that are insoluble in toluene or benzene) condensation product as indicated from oxidation where the scheme of reaction (*66*, *67*) is:

$$\text{oils} \rightarrow \text{resins} \rightarrow \text{asphaltenes}$$

High aromaticity is generally prevalent in the asphaltenes and the resins, but in some resins the hydrocarbons show an increase in aliphatic material (more sidechains, for example) until, with considerable saturation, the oils, which contain numerous alkyl chains of varying length, are reached. The degree of aromaticity is important when the resins are being adsorbed by the asphaltene particles, just as paraffinicity is important when the resins are desorbed. High aromaticity of the maltenes (i.e., that part of petroleum remaining after the asphaltenes have been removed, often referred to as the deasphalted oil) indicates good solvency for the asphaltenes, and the solvent power of the maltenes is one of the most important factors in determining the physico-chemical behavior of the petroleum colloid system.

The means by which the asphaltenes remain dispersed in the oil medium [asphaltenes are, in fact, insoluble in the oil fraction (*5*)] has been the subject of much speculation, but it is now clear that this is mainly attributable to the resins. In fact, it is possible to bring about dispersion of asphaltenes in their corresponding oil fractions only by addition of resins. The general indications are that the degree of aromaticity of, and the proportion of heteroelements in, the resins appear to play an important part in the ability of these materials to bring about solubility of the asphaltenes in an oil. Furthermore, if the added resins are less aromatic than the natural resins, asphaltenes will not disperse; for resins of equal aromaticity, those having higher heteroelement content are better for dispersion purposes. Therefore, it appears that a crude oil is a complex system within each fraction, dependent upon other systems for complete mobility and solubility (*5*). It is presumed that the resins associate with the asphaltenes in the manner of an electron donor–acceptor, and that there could well be several points of structural similarity between the asphaltenes and resins that would have an adverse effect on the ability of the resins to associate with asphaltenes from a different crude oil (*5, 43*).

An early hypothesis of the physical structure of petroleum (*52*) indicated that asphaltenes are the centers of micelles formed by adsorption, or even by absorption of part of the maltenes, that is, resin material, onto the surfaces or into the interiors of the asphaltene particles. Thus, most of those substances with greater molecular weight and with the most pronounced aromatic nature are situated closest to the nucleus and are surrounded by lighter constituents of less aromatic nature. The transition of the intermicellular (dispersed or oil) phase is gradual and almost continuous. Continued attention to this aspect of asphaltene chemistry has led to the assumption that asphaltenes exist as clusters within the micelle. This arises mainly because of the tendency for asphaltenes to associate in dilute solution in solvents of low polarity and from possible misinterpretation of viscosity data (*58, 64*). The presence of asphaltene "stacks" in the solid phase, as deduced from x-ray diffraction patterns (*68*), also seemed to support the concept of the widespread existence of asphaltene clusters in the micelle.

The concept of hydrogen-bonding interactions being one of the means of association between the asphaltenes and resins has, however, led to a reconsideration of the assumed cluster as part of the micelle (*43*). Indeed, it appears that when resins and asphaltenes are present together, hydrogen-bonding may be one of the mechanisms by which resin–asphaltene interactions are achieved. Resin–asphaltene interactions are preferred over asphaltene–asphaltene interactions (*43*). Thus, if the same intermolecular forces are present in petroleums and bitumens, it would perhaps not be surprising that asphaltenes exist not as the more cumbersome agglomerations but more nearly single entities that are peptized, and effectively dispersed, by the resins.

Literature Cited

1. Speight, J. G. *Appl. Spectrosc. Rev.* **1972,** *5*, 211.
2. Yen, T. F. *Am. Chem. Soc., Div. Pet. Chem., Prepr.* (New York, Aug.–Sept., 1972) *17*(4), F102.
3. Yen, T. F. *Fuel* **1970,** *49*, 134.

4. Yen, T. F. *Am. Chem. Soc., Div. Fuel Chem., Prepr.* (Los Angeles, Mar.–Apr., 1971) *15*(1), 57.
5. Koots, J. A.; Speight, J. G. *Fuel* **1975,** *54,* 179.
6. Mitchell, D. L.; Speight, J. G. *Fuel* **1973,** *52,* 149.
7. Bland, W. F.; Davidson, R. L., Eds. "Petroleum Processing Handbook"; McGraw–Hill: New York, 1967; p. 3.
8. *Am. Soc. Test. Mater., Book ASTM Stand.* **1975,** *24.*
9. *Am. Soc. Test. Mater., Book ASTM Stand.* **1975,** *15.*
10. "Standards for Petroleum and Its Products, Standard No. 1P 143/57"; Institute of Petroleum: London.
11. McKay, J. F.; Amend, P. J.; Cogswell, T. E.; Hornsberger, P. M.; Erickson, R. B.; Latham, D. R. *Am. Chem. Soc., Div. Pet. Chem., Prepr.* (New Orleans, Mar., 1977) *22*(2), 708.
12. Speight, J. G., unpublished data, 1969.
13. Speight, J. G. "The Structure of Petroleum Asphaltenes," In "Information Series"; Alberta Research Council: 1978; No. 81.
14. Speight, J. G. *178th Nat. Meet. Am. Chem. Soc., Washington, D.C., Sept. 1979.*
15. Yen, T. F.; Erdman, J. G.; Pollack, S. S. *Anal. Chem.* **1961,** *33,* 1587.
16. Speight, J. G., presented at *U. S. Bur. Mines, Symp. Fossil Chem. and Energy, Laramie, WY, July 23–27, 1974.*
17. Speight, J. G. *Proc. Nat. Sci. Found. Symp. Fund. Org. Chem. Coal, Knoxville, TN, 1975, 125.*
18. Speight, J. G., unpublished data, 1971.
19. Ibid., 1977–1978.
20. Speight, J. G. *Fuel* **1970,** *49,* 76.
21. Ibid., **1971,** *50,* 102.
22. Ibid., **1970,** *49,* 134.
23. Speight, J. G. *Am. Chem. Soc., Div. Fuel Chem., Prepr.* (Los Angeles, Mar.–Apr., 1971) *15*(1), 57.
24. Ritchie, R. G. S.; Roche, R. S.; Steedman, W. *Fuel* **1979,** *58,* 523.
25. Ignasiak, T.; Kemp-Jones, A. V.; Strausz, O. P. *Org. Chem.* **1977,** *42,* 312.
26. Moschopedis, S. E.; Speight, J. G., unpublished data, 1977.
27. Holy, N. L. *Chem. Rev.* **1974,** *74,* 243.
28. Larsen, J. W.; Urban, L. O. *J. Org. Chem.* **1979,** *44,* 3219.
29. Speight, J. G.; Moschopedis, S. E. *Fuel,* **1980,** *59,* 440.
30. Ignasiak, T.; Strausz, O. P. *Fuel* **1978,** *57,* 617.
31. Benjamin, B. M. *Fuel* **1978,** *57,* 378.
32. Hooper, R. J.; Battaerd, H. A. J.; Evans, D. G. *Fuel* **1979,** *58,* 132.
33. Clerc, R. J.; O'Neal, M. J. *Anal. Chem.* **1961,** *33,* 380.
34. Moschopedis, S. E.; Parkash, S.; Speight, J. G. *Fuel* **1978,** *57,* 431.
35. Speight, J. G.; Penzes, S. *Chem. Ind. (London)* **1978,** 729.
36. Kharasch, N.; Meyers, C. Y. "The Chemistry of Organic Sulphur Compounds"; Pergamon: New York, 1966; Vol. 12.
37. Speight, J. G., Ph.D. Thesis, Univ. of Manchester, 1965.
38. Speight, J. G., unpublished data, 1977.
39. Druschel, H. V. *Am. Chem. Soc., Div. Petrol. Chem., Prepr.* (New York, Aug.–Sept., 1972) *17*(4), F92.
40. Knotnerus, J. *Am. Chem. Soc., Div. Petrol. Chem., Prepr.* (Los Angeles, Mar.–Apr. 1971) *16*(1), D37.
41. Petersen, J. C.; Barbour, F. A.; Dorrence, S. M. *Proc. Assoc. Asphalt Paving Technol.* **1974,** *43,* 162.
42. Moschopedis, S. E.; Speight, J. G. *Am. Chem. Soc., Div. Fuel Chem., Prepr.* (San Francisco, Aug.–Sept., 1976) *21*(6), 198.
43. Moschopedis, S. E.; Speight, J. G. *Fuel* **1976,** *55,* 187.
44. Ibid., 334.
45. Nicksic, S. W.; Jeffries-Harris, M. J. *J. Inst. Pet., London* **1968,** *54,* 107.
46. Moschopedis, S. E.; Speight, J. G. *Am. Chem. Soc., Div. Pet. Chem., Prepr.* (Washington, D.C., Sept., 1979) *24*(4), 1007.
47. Winniford, R. S. *Inst. of Pet. Rev.* **1963,** *49,* 215.

48. Speight, J. G.; Moschopedis, S. E. *Fuel* **1977,** *56,* 344.
49. Ray, B. R.; Witherspoon, P. A.; Grim, R. E. *Phys. Chem.* **1957,** *61,* 1296.
50. Wales, M.; van der Waarden, M. *Am. Chem. Soc., Div. Pet. Chem., Prepr.* (Philadelphia, Apr., 1964) *9*(2), B-21.
51. Labout, J. W. A. "Properties of Asphaltic Bitumen"; Pfeiffer, J. P., Ed.; Elsevier: New York, 1950; p. 35.
52. Pfeiffer, J. P.; Saal, R. N. *Phys. Chem.* **1940,** *44,* 139.
53. Griffin, R. L.; Simpson, W. C.; Miles, T. K. *Chem. Eng. Data* **1959,** *4*(4), 349.
54. Hillman, E.; Barnett, B. *Proc. 4th Ann. Meet. ASTM* **1937,** *37*(2), 558.
55. Sakhanov, A.; Vassiliev, N. *Petrol. Zeit.* **1927,** *23,* 1618.
56. Katz, M. *Can. J. Res.* **1934,** *10,* 435.
57. Grader, R. *Oel Kohle* **1942,** *38,* 867.
58. Mack, C. *Phys. Chem.* **1932,** *36,* 2901.
59. Reerink, H. *Ind. Eng. Chem., Prod. Res. Dev.* **1973,** *12,* 82.
60. Markhasin, I. L.; Svirskaya, O. D.; Strads, L. N. *Kolloid-Z.* **1969,** *31,* 299.
61. Altgelt, K. H. *Am. Chem. Soc., Div. Pet. Chem., Prepr.* (Atlantic City, Sept., 1968) *13*(3), 37.
62. Kirby, W. *Soc. Chem. Ind.* **1943,** *62,* 58.
63. Lerer, M. *Ann. Comb. Liqu.* **1934,** *9,* 511.
64. Moschopedis, S. E.; Fryer, J. F.; Speight, J. G. *Fuel* **1976,** *55,* 227.
65. Ignasiak, T.; Strausz, O. P.; Montgomery, D. S. *Fuel* **1977,** *56,* 359.
66. Moschopedis, S. E.; Speight, J. G. *Proc. Assoc. Asphalt Paving Technol.* **1976,** *45,* 78.
67. Moschopedis, S. E.; Speight, J. G. *J. Mater. Sci.* **1977,** *12,* 990.
68. Dickie, J. P.; Yen, T. F. *Anal. Chem.* **1967,** *39,* 1847.

RECEIVED June 23, 1980.

2

The Concept of Asphaltenes

ROBERT B. LONG

Corporate Research—Science Laboratories, Exxon Research and Engineering Company, P.O. Box 45, Linden, NJ 07036

This chapter considers the composition of heavy oils in terms of a map of molecular weight vs. polarity of the various components in the sample. Thus, a clearer understanding of the nature of asphaltenes is obtained. The inter-relationship of polarity and molecular weight in terms of solubility behavior can be better understood and it becomes clear that there is not a specific chemical composition or a specific molecular weight description for asphaltenes. Rather, asphaltenes contain a wide distribution of polarities and molecular weights.

The classic definition of asphaltenes is based on the solution properties of petroleum residuum in various solvents. This generalized concept has been extended to fractions derived from other carbonaceous sources, such as coal and oil shale. With this extension there has been much effort to define asphaltenes in terms of chemical structure and elemental analysis as well as by the carbonaceous source. This effort is summarized by Speight and Moschopedis (*1*) in their chapter in this volume along with a good summary of the current thinking. Thus, there are "petroleum asphaltenes," "coal tar asphaltenes," "shale oil asphaltenes," "tar sands bitumen asphaltenes," and so on. In this chapter I will attempt to show how these materials are special cases of an overall concept based directly on the physical chemistry of solutions and that the idea that they have a specific chemical composition and molecular weight is incorrect even for different crude oil sources.

Definitions of Asphaltenes

In Figure 1, the classic definition of asphaltenes is illustrated. This definition is an operational one; that is, asphaltenes are soluble in benzene and insoluble in pentane. Usually, for virgin petroleum samples, the residuum is completely soluble in benzene. However, with heat-soaked samples or coal-derived liquids, the benzene insolubles can be appreciable. Therefore, further

0065-2393/81/0195-0017$05.00/0

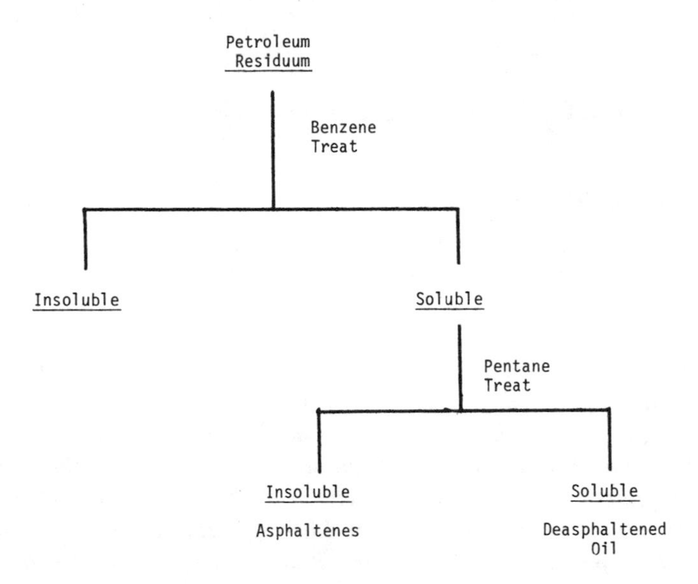

Figure 1. Separation of asphaltenes from petroleum residuum

classifications such as mesophase (*2*) and asphaltols (*3*) have been made for the benzene insolubles based on solubility or insolubility in more potent solvents such as pyridine and quinoline.

In addition to the further classification of benzene insolubles, there also has been a growing tendency to classify asphaltenes by the particular paraffin used to precipitate them from the benzene-soluble portion of the feed. Thus, there are pentane asphaltenes, hexane asphaltenes, heptane asphaltenes, and so on. However, the present tendency is to define the material precipitated by *n*-heptane as asphaltenes. Corbett and Petrossi (*4*) have summarized the present state of the art in excellent fashion and Figure 2, taken from their paper, shows how the yield of paraffin insolubles changes with the carbon number of the precipitating solvent for Arab light atmospheric residuum. There is very little difference in the amount precipitated for *n*-heptane and heavier *n*-paraffins. This means that only the most insoluble materials in the crude are being precipitated by these solvents. However, as the precipitating *n*-paraffin molecule gets smaller, marked increases in the amount of precipitate occur. Thus, something in addition to *n*-heptane asphaltenes is coming out of solution in these latter cases. Comparison of pentane precipitation with heptane precipitation in Figure 2 shows that about three times as much precipitate is recovered with pentane. The material representing this difference has a different composition and molecular weight from the *n*-heptane asphaltenes.

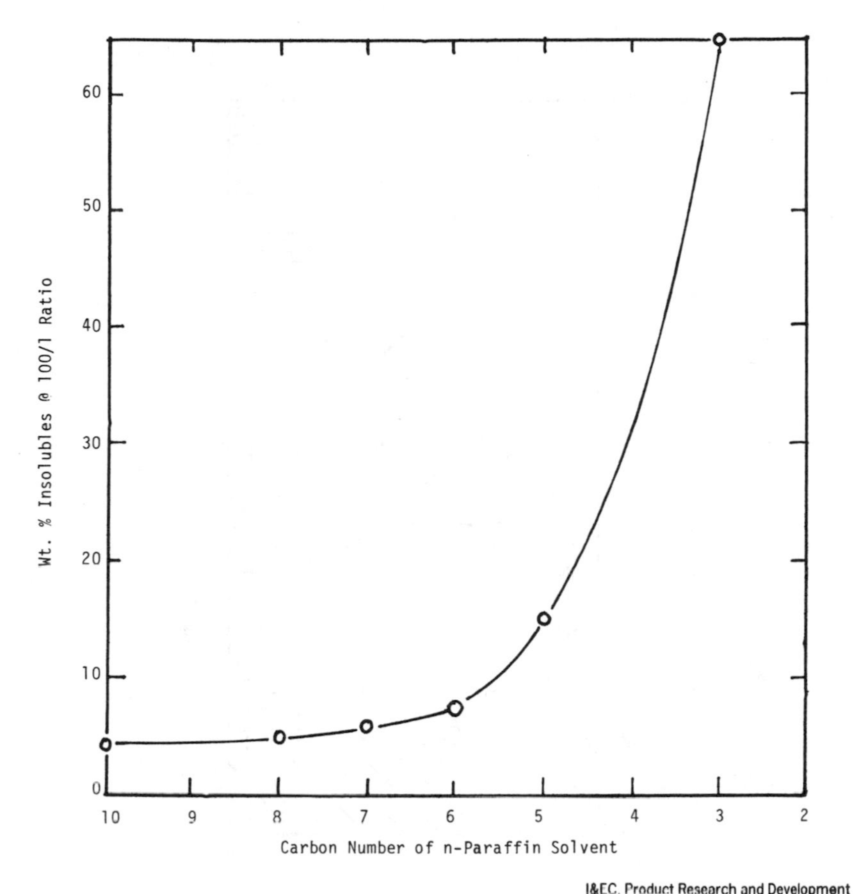

I&EC, Product Research and Development

Figure 2. Effect of solvent carbon number on insolubles (Arabian Light atmospheric residuum) (4)

Composition of Asphaltenes

In the same paper, Corbett and Petrossi (*4*) show the distribution of molecular types and sizes in a typical residuum. This distribution is based on a generic fractionation using solvent deasphaltening and elution–adsorption chromatography (5) and is reproduced as Figure 3. The upper portion of the chart is obtained by elution–adsorption separations on distillate fractions and shows that the proportion of saturates decreases as the boiling point increases, while the amount of polar aromatics increases as the boiling point increases. The naphthene-aromatic content stays relatively constant. The end point of the vacuum distillation in this case is 565°C atmospheric equivalent boiling point and the molecular weight of the heaviest distillate is only 552. The portion below this line is a vacuum residuum and is fractionated by propane,

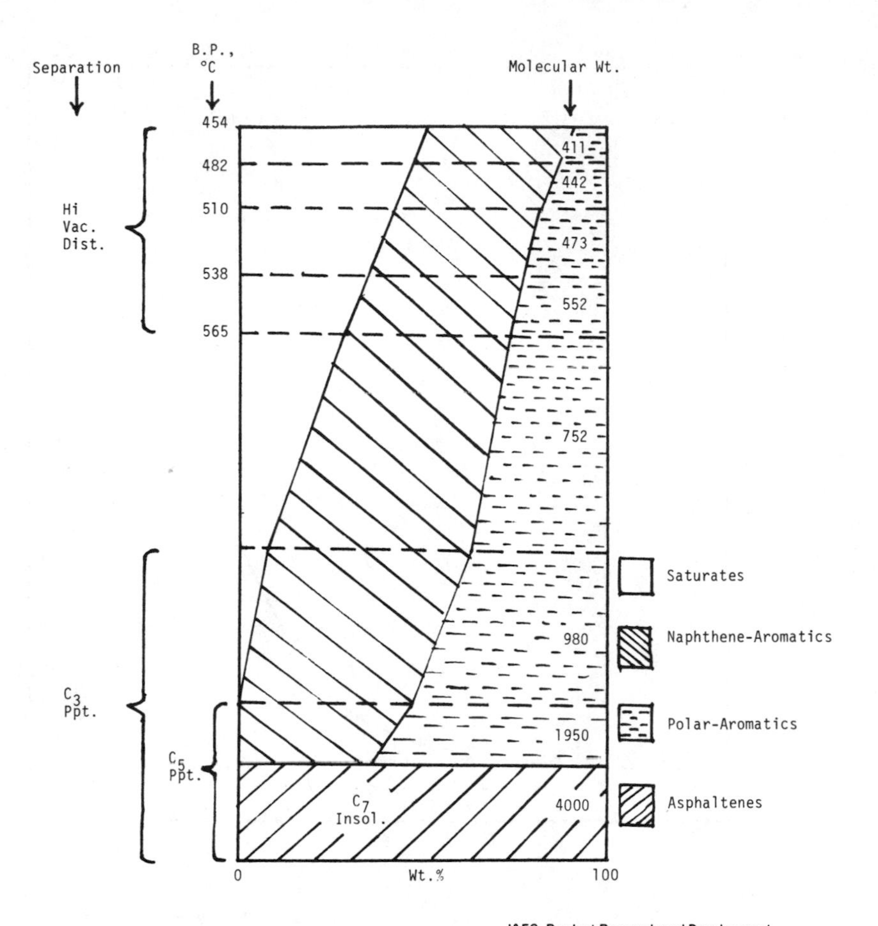

I&EC, Product Research and Development

Figure 3. Distribution of molecular types and sizes in a residuum (4)

n-pentane, and *n*-heptane precipitation. The *n*-heptane precipitate is taken as asphaltenes by definition. However, more recent work in our laboratories has shown that the *n*-heptane asphaltenes consist primarily of polar aromatics with an appreciable content of naphthene-aromatics, and traces of saturates. This would be expected if the trend shown in Figure 3 for saturates, naphthene aromatics, and polar aromatics were continued through the asphaltene fraction.

It is also interesting to note the dramatic increase in molecular weight of the fractions as the bottom of the figure is approached.

The New Concept

Based on the above types of information and solubility theory for nonelectrolytes as given in Hildebrand and Scott (*6*), a new way of describing asphaltenes has emerged. This description is illustrated in Figure 4. It is known intuitively by experts in asphalt composition but, to our knowledge, has not been described in this way before. However, McKay et al. (*7*) have recognized that asphaltenes include the most polar and the highest molecular weight species in petroleum. Figure 4 shows a rectangular plot of molecular weight vs. polarity for the molecular types found in petroleum residuum. The polarity scale is in definable arbitrary units, such as relative adsorptive strength on a solid such as Attapulgus clay and/or silica gel as in the elution-adsorption technique of Corbett (*5*), or by solubility in a variety of solvents of increasing polarity as practiced on separation of coal-liquid fractions by Farcasiu et al. (*3*). The point is that this scale is a molecular-weight-independent portion of the solubility parameter. The diagonal line shown for *n*-heptane is drawn so that the material above and to the right of it is representative of the *n*-heptane asphaltene fraction. That is, less polar materials of higher molecular weight and more polar materials of lower molecular weight both precipitate as asphaltenes. As we move to *n*-pentane as the precipitative agent, both less polar and lower molecular weight materials are included in the precipitate and the total amount of precipitate increases in agreement with Corbett and Petrossi's results (*4*). The shape of the pentane and heptane precipitation lines, shown here as straight lines, should depend on the scale used for characterizing polarity. Furthermore, the shape of the top and right-hand boundaries of the field of the graph depend on what is actually in the residuum. From the data shown in Figure 3, we would expect the top boundary in molecular weight for low polarity fractions to be lower than for high polarity fractions. This would lead to a sloping line of maximum molecular weight with its lower end at low polarity and its upper end at high polarity. This is illustrated by lines B, C, and D in Figure 5. The shape of the particular top boundary line is a characteristic of the crude source.

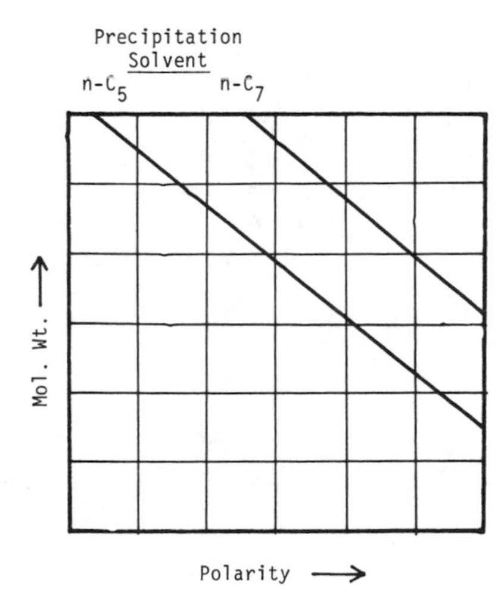

Figure 4. Petroleum asphaltene precipitation

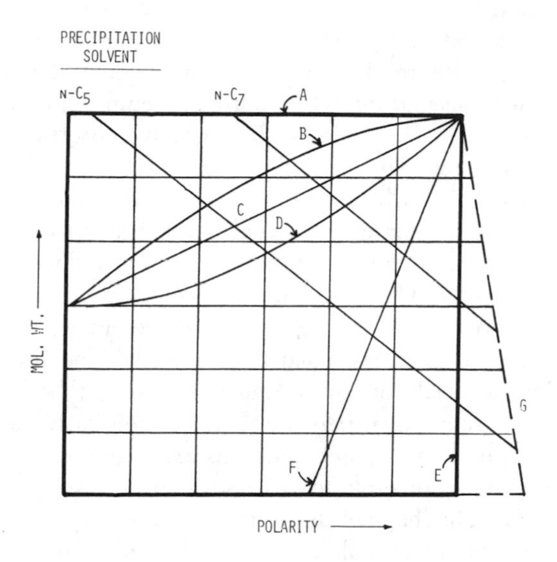

Figure 5. Petroleum asphaltene precipitation

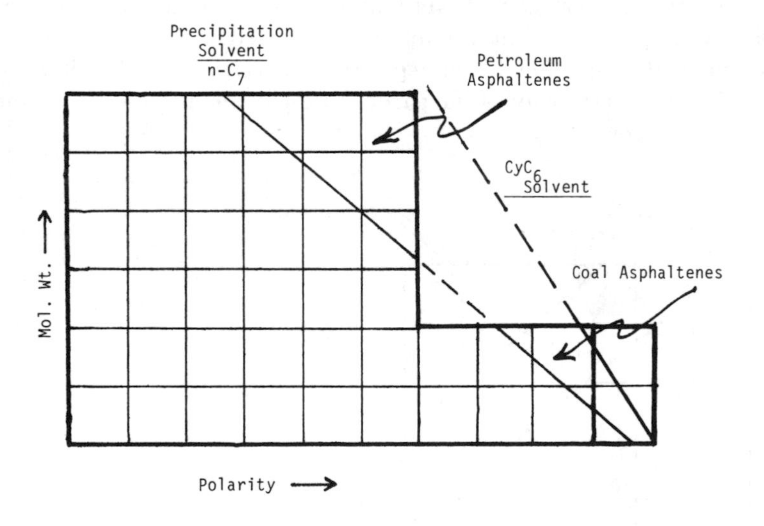

Figure 6. Comparison of petroleum and coal asphaltenes

Line A is the most unlikely with lines B and D being more likely than line C. Similarly, boundary line E in Figure 5 is also unlikely, but we do not yet have enough data to see if either line F or G is more general.

An extension of this treatment to coal liquids is shown in Figure 6. Here, the coal liquids are shown as much lower in molecular weight than the petroleum residuum and also, because of high content of phenol and other oxygenated compounds, extending to much higher polarities. An extension of the *n*-heptane precipitation line into this region shows that material that would be called coal-liquids asphaltenes by our definition should have lower molecular weight and higher polarity than asphaltenes from petroleum. Again, the boundaries of the compositional field will depend on the source of the coal liquids and their method of preparation.

In Table I, a comparison is made of the elemental composition of typical asphaltenes from petroleum and coal liquids. This table shows the typical lower H/C ratio and higher oxygen content for the coal asphaltenes. Furthermore, the GPC molecular-weight distributions shown in Figure 7 illustrate the higher molecular-weight of petroleum asphaltenes as well as the wider molecular-weight distribution.

In Table II, the elemental composition of *n*-heptane asphaltenes from a number of crude sources is shown. The range of composition indicates a range of polarities is quite likely since polarity generally varies largely with the heteroatom content of the fraction. This is particularly true for oxygen and nitrogen content.

Finally, Figure 8 illustrates the molecular-weight distributions obtained by GPC of a number of *n*-heptane asphaltenes from various crude sources. As can be seen, a wide range of molecular weights can be obtained for asphaltenes from different crude sources. Furthermore, the peak in the distribution occurs at different values for the different crudes. It is well known that molecular association of asphaltene molecules can be a problem in molecular weight determination either by GPC or VPO. However, the real extent of this problem is quite problematical. Small angle x-ray scattering

Table I. Comparison of Coal and Petroleum Asphaltenes

Inspection	*Coal Asphaltenes*	*Petroleum Asphaltenes*
Carbon (wt %)	86.93	81.7
Hydrogen (wt %)	6.83	7.60
Nitrogen (wt %)	1.36	1.23
Sulfur (wt %)	1.09	7.72
Oxygen (wt %)	3.8	1.7
Vanadium (ppm)	9	1200
Nickel (ppm)	3	390
VPO, molecular weight	726	5400
H/C	0.94	1.12

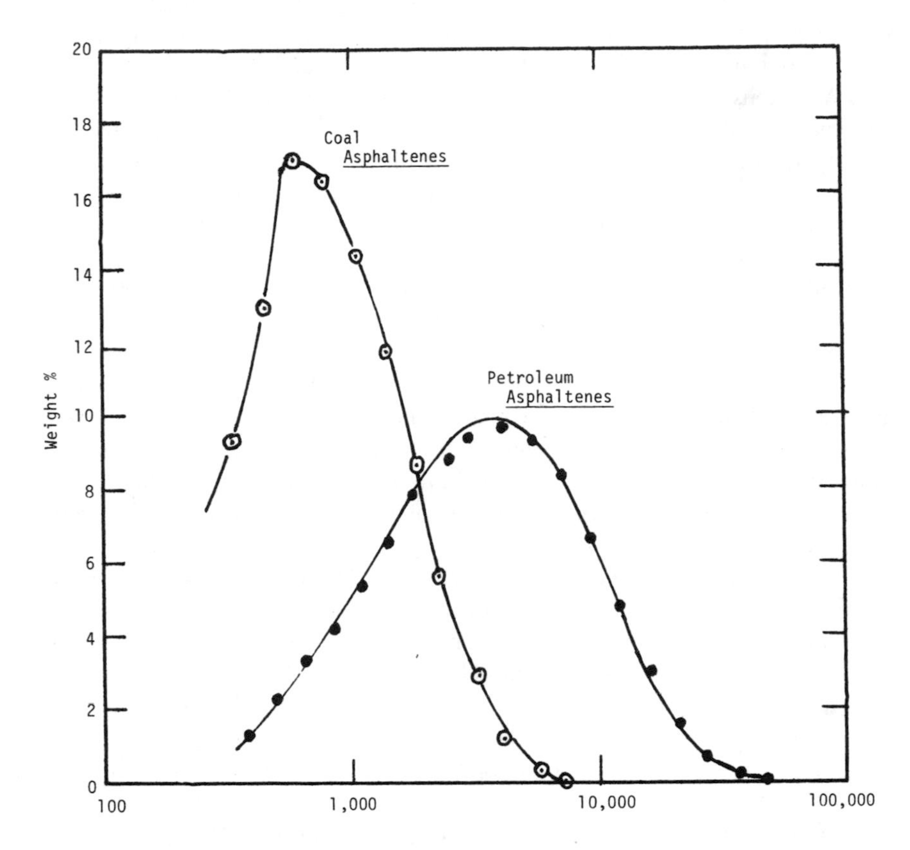

Figure 7. Comparison of molecular weight distribution of (⊙) coal asphaltenes and Jobo petroleum asphaltenes: (——) vacuum residuum and (●) atmospheric residuum

studies (8) have shown that for Jobo asphaltenes the size of the asphaltene micelle was independent of both concentration and the nature of the solvent, while for Tia Juana asphaltenes the nature of the solvent had a big effect on size. Thus, it appears that the Tia Juana asphaltenes may be associated much more than the Jobo asphaltenes.

Because of these considerations, molecular weights determined by GPC may be too high but are probably relative. Furthermore, there is no correct molecular weight for asphaltenes, but a distribution as shown in Figure 8. Thus, these data tend to support some type of generalized concept as given in Figures 4, 5, and 6.

To quantify the concepts given above, it is necessary to devise a polarity scale related to solubility theory. Since the precipitating solvents for asphaltene separation are at the low end of the solubility parameter scale, it seems

Table II. Composition of Asphaltenes from Different Crude Sources

	Venezuela			
Inspection	*Tia Juana*	*Jobo*	*Cold Lake*	*Arabian Heavy*
Carbon (wt %)	85.04	83.21	82.04	82.67
Hydrogen (wt %)	7.68	7.89	7.90	7.64
Nitrogen (wt %)	1.33	1.86	1.21	1.00
Sulfur (wt %)	3.96	5.63	7.72	7.85
Oxygen (wt %)	1.8[a]	1.68	1.7	0.89
Vanadium (ppm)	2000	2960	1200	720
Nickel (ppm)	255	468	392	176
H/C	1.09	1.14	1.16	1.11

[a]By difference.

likely that some form of polarity scale based on the difference in solubility parameter between the resid fractions and *n*-heptane should produce satisfactory correlations.

While the above treatment lays out in a generalized way the concept of asphaltenes, there are some specific additional effects that bear on the precipitation of a given type of molecule from hydrocarbon solution. For

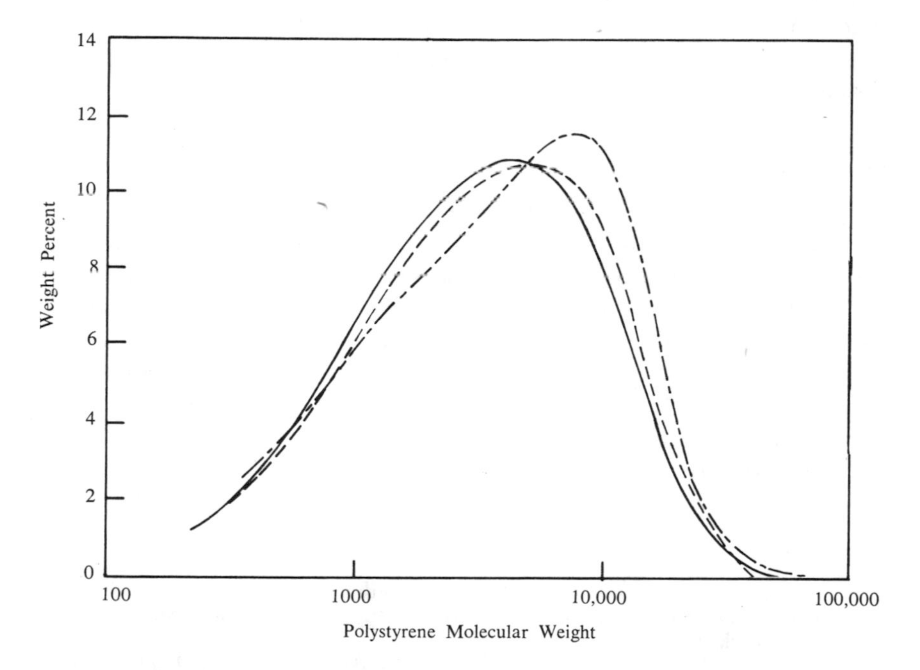

Figure 8. Comparison of asphaltenes (n – C_7) from various crudes: (——) *Cold Lake,* (---) *Arabian Heavy,* (-·-) *Tia Juana Medium*

example, the crystal lattice energy for high-melting solids must affect precipitation behavior. Asphalt chemists have known for a long time that some crudes precipitate high-melting waxes in their asphaltenes. Drushel (*9*) has found that by performing the deasphaltening at a temperature above the melting point of the waxes in the feed, the waxes do not precipitate. Thus, the lattice energy of the wax crystals must be sufficient to cause precipitation at lower temperatures, whereas lower melting paraffinic molecules stay in the heptane solution. I believe similar effects are operative for the polynuclear aromatics largely denuded of alkyl groups that are often found in coal-derived liquids.

In addition to the effects of crystal lattice energy, choice of solvent other than *n*-paraffins can also be important. For example, Corbett and Swarbrick (*10*) have shown that a number of oxygenated compounds can precipitate asphaltenes from petroleum residua in quantities varying from 12 wt % to 100 wt % on resid. Clearly, the shape of the precipitation curves for the different solvents should be different from that for *n*-paraffins and from each other. However, the use of a solubility parameter type of polarity scale should permit rationalization of the results when considered along with the solubility parameter of the particular solvent.

An additional subtlety in solvent choice is shown by cyclohexane and methylcyclohexane. In these cases, petroleum asphaltenes are completely soluble, while coal-derived asphaltenes are only sparingly soluble. This would imply that the precipitation line for cyclohexane runs completely outside the petroleum field of Figures 5 and 6 but passes through the coal-liquids field in Figure 6. Thus, some care must be shown when considering the equivalence or nonequivalence of various solvents.

Conclusion

The concept of asphaltenes is rooted in the solubility behavior of high-boiling hydrocarbonaceous materials in benzene and low-molecular-weight *n*-paraffin hydrocarbons. This behavior is a result of physical chemistry effects that are caused by a spectrum of chemical properties. This chapter has pointed out that by considering molecular weight and molecular polarity as separate properties of molecules, the solvent-precipitation behavior of materials derived from various carbonaceous sources can be understood. Future quantification of this approach probably can be achieved by developing a polarity scale based on solubility parameter.

Literature Cited

1. Speight, J. G.; Moschopedis, S. E. Chapter 1 in this book.
2. Brooks, J. D.; Taylor, G. H. "Chemistry & Physics of Carbon"; Marcel Decker: New York, 1968; Vol. 4, p. 243.
3. Farcasiu, M.; Mitchell, T. O.; Whitehurst, D. D. *Am. Chem. Soc., Div. Fuel Chem., Prepr.* (San Francisco, Aug.–Sept., 1976) *21*(7), 11.
4. Corbett, L. W.; Petrossi, U. *Ind. Eng. Chem., Prod. Res. Dev.* **1978,** *17*, 342.

5. Corbett, L. W. *Anal. Chem.* **1969,** *41,* 576.
6. Hildebrand, J. H.; Scott, R. L. "The Solubility of Non-Electrolytes"; Dover Publications, Inc.: New York, 1964.
7. McKay, J. F.; Amend, P. J.; Cogswell, T. E.; Harnsberger, P. M.; Erickson, R. B.; Latham, D. R. In "Petroleum Asphaltenes: Chemistry and Composition", *Adv. Chem. Ser.* **1978,** *170,* 128–142.
8. Kim, H.; Long, R. B. *Ind. Eng. Chem. Fundam.* **1979,** *18,*(1), 60.
9. Drushel, H. V., private communication.
10. Corbett, L. W.; Swarbrick, R. E., private communication.

RECEIVED June 23, 1980.

3

Coal-Derived Asphaltenes

History and Some Current Observations

B. C. BOCKRATH

U. S. Department of Energy, Pittsburgh Energy Technology Center, P. O. Box 10940, Pittsburgh, PA 15236

F. K. SCHWEIGHARDT

Air Products and Chemicals, Inc., Corporate Research and Development Division, P. O. Box 538, Allentown, PA 18105

A historical overview of research on the isolation, separation, and characterization of asphaltenes found in coal-derived liquids is given. Compared with asphaltenes isolated from crude petroleum, coal-derived asphaltenes are generally lower in molecular weight and in hydrogen and sulfur contents but are of higher aromaticity. The relationship of phenolic and ether groups to the conversion and solubility of coal-derived asphaltenes and the influence of hydrogen bonding on the viscosity of coal-derived liquids is discussed. As conversion of asphaltenes to pentane-soluble oils by hydrotreatment increases, their H/C and O/C ratios decline and their aromaticity increases. The viscosity of coal-derived liquids depends in part upon the asphaltene content of the liquid and also on the phenol content and molecular weight of the asphaltenes.

The substances called asphaltenes are important intermediates in the hydrogenation of coals in the liquid phase. How the word asphaltene originated and how these substances have been characterized are the subjects of the first part of this chapter.

Discovery of Asphaltenes

The word asphaltene was coined in France by J. B. Boussingault (*1*) in 1837. Boussingault described the constituents of a number of bitumens (asphalts) found at that time in eastern France and in Peru. After separation of the components of an asphalt by distillation, he named the volatile oily constituents that were ether soluble, *petrolenes*. He named the alcohol-insoluble, essence of turpentine soluble solids obtained from the distillation residue *asphaltene*, since it resembled the original asphalt. He concluded from his findings that the ratio of these two principals composing bitumen varies continuously from source to source, thereby giving each bitumen a different degree of fluidity.

In modern terms, asphaltene is conceptually defined as the *n*-pentane-insoluble and benzene-soluble fraction whether it is derived from coal or from petroleum. There are a number of procedures used to isolate asphaltene (*2–7*), all of which appear to be reproducible (*8*) but do not necessarily provide equivalent end-products. The similarity between coal- and petroleum-derived asphaltenes begins and ends at the definition of the separation procedure. Puzinauskas and Corbett's (*9*) comments on asphalt may be paraphrased and applied to asphaltene. They state that the broad solvent classification is unfortunate; it leads to misconceptions that petroleum and coal materials are alike, or at least similar. However, these two classes of materials differ not only in their origin, mode of manufacture and uses, but also in their chemical composition and physical behavior.

We believe it is important, at this time, to establish a documented historical review of asphaltenes and the separation/characterization procedures used by early petroleum and coal chemists to profile crude oils and the products of coal hydrogenation. We shall explore the most important differences between petroleum- and coal-derived asphaltenes.

Petroleum- Versus Coal-Derived Asphaltenes

Table I summarizes comparisons between asphaltenes derived from bituminous coal liquefaction and those derived from petroleum crudes. The molecular size and atomic H/C ratios suggest a molecular profile quite different for the two asphaltenes. The ranges represent, as best as could be found, reasonable extremes for each of the properties. We are well aware that the number-average molecular weight of petroleum asphaltenes has been influenced by aggregate formation. To overcome this effect, molecular weight determinations should be made in dilute noninteracting solvents (e.g., methylene chloride), and solutions should be filtered or ultracentrifuged in helium-degassed solvents.

Benzene solubility of petroleum asphaltenes apparently is attributable to the structural arrangement of the long aliphatic sidechains on aromatic rings. Coal-derived asphaltenes are also soluble in benzene, but by comparison are of smaller size and molecular weight and have a greater concentration of functional groups. The nature of the functional groups also appears to be different, for example, oxygen appears as hydroxyl and ether moieties in coal-derived asphaltenes, while petroleum asphaltenes contain hydroxyl, carbonyl, and some ether linkages. Nitrogen, as well as oxygen, has only two

Table I. Comparisons of Properties Between Petroleum- and Coal-Derived Asphaltenes

Property	*Asphaltenes* *Coal*	*Asphaltenes* *Petroleum*
Solubility (benzene soluble/pentane insoluble)	yes	yes
Number-average molecular weight	400–800	2000–20,000
Weight percent carbon	84–89	80–86
Weight percent hydrogen	5.8–7.3	7.7–9.3
Weight percent oxygen	2–6	1–2.7
Weight percent nitrogen	1–3	0.3–0.5
Weight percent sulfur	0.1–2	3–9.3
Percent ash by low or high temperature	<0.1	<2
Aromaticity, f_a, by ^{13}C NMR	0.6–0.7	0.4–0.55
Forms of oxygen	OH,—O—	OH,—C=O,—O—
Forms of nitrogen	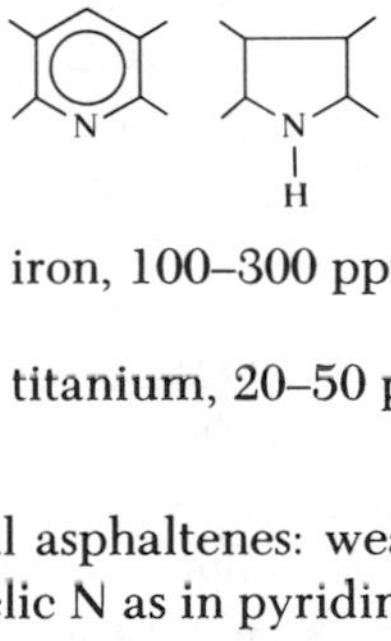 [pyridine structure, N] [pyrrole structure, N–H]	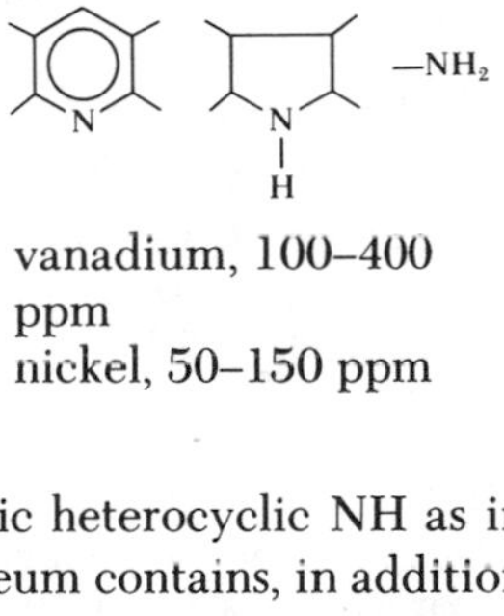 [pyridine structure, N] [pyrrole structure, N–H] $—NH_2$
Metals	iron, 100–300 ppm; titanium, 20–50 ppm	vanadium, 100–400 ppm; nickel, 50–150 ppm

predominant forms in coal asphaltenes: weakly acidic heterocyclic NH as in pyrole and basic heterocyclic N as in pyridine. Petroleum contains, in addition to those groups, NH_2 functions.

Figure 1 shows the atomic H/C vs. O/C ratios for a series of fossil fuels. It is evident from the figure that coal-derived asphaltenes stand alone, with H/C atomic ratio about 60% of those derived from other fossil fuels. The asphaltenes from coal liquefaction products have experienced hydrotreatment at high pressures (2000–4000 psi) and temperatures near 450° C, yet they are much more aromatic than petroleum asphaltenes. Asphaltenes from coal have an aromaticity, f_a, determined from ^{13}C nuclear magnetic resonance, of 0.60–0.70, while petroleum values are listed as 0.40–0.55.

Early Attempts at Characterization

Marcusson (*10*) was one of the first in this century to observe similarities between the behavior of asphaltenes of crude oils and native asphalts and coal.

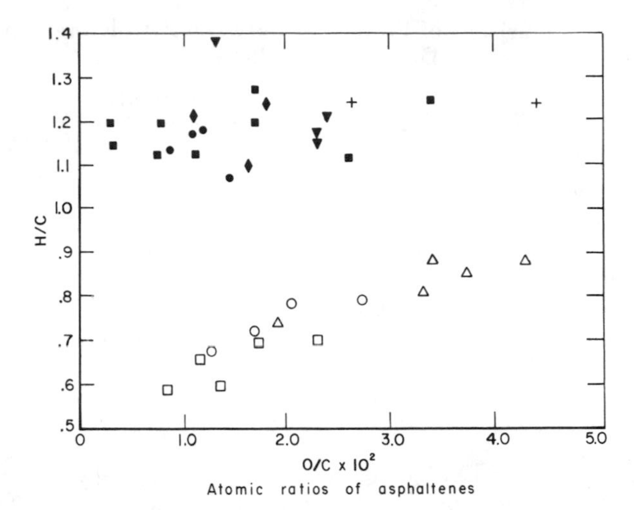

Figure 1. Atomic H/C vs. O/C ratios for asphaltenes from a variety of sources: (■) petroleum crude (Alberta, Cretaceous, Carboniferous, and Devonian); (●) Middle East crude; (▼) North and South American crude; (+) shale oil; (♦) oil sands; and (Δ) liquids from Kentucky coal at 4000 psi hydrogen and West Virginia coal at (○) 4000 psi, and (□) at 2000 psi hydrogen (21)

He reported taking powdered coal and asphaltenes from petroleum, covering them with ether and passing hydrogen chloride gas through the suspension or solution. Marcusson observed that coal picked up 3.5% chlorine, while petroleum asphaltenes had 2.5% chlorine, and that similar results could not be obtained with aqueous hydrogen chloride. Finally he observed that coal in the presence of hydrogen iodide in sealed tubes formed petroleum-like hydrocarbons, similar to the earlier work of Berthelot (*11*).

An early characterization of asphaltenes separated from petroleum by their solubility in carbon tetrachloride was put forth by Nellensteyn in 1930 (*12*). He described asphaltenes as colloidal systems as opposed to being hydrocarbons in mutual solution with petrolenes/maltenes. According to Nellensteyn, asphaltenes were dispersed carbon or high molecular weight hydrocarbons, absorbed on the surface of low molecular weight hydrocarbons. At this time it appeared to many investigators that the molecular composition of the different constituents was less important than the relative distribution of the constituents. These thoughts were similar to those of Boussingault (*1*) almost 100 years earlier.

Early Chromatographic Separations

About the same time (1930s) that analytical methods based on solubility (*13*) were refined, comparisons were made with separations based on newly developed methods of adsorption chromatography (*14*) for the separation of petroleum asphaltenes and later for the products of coal hydrogenation. Much of this early work took place in Germany, especially at Ludwigshafen.

Poell's method (*13*) for separating coal hydrogenation products used elution chromatography with petroleum ether over Fuller's earth, successive elution with chloroform to free the oils, and finally pyridine to isolate asphaltenes. It was reported that Poell's method (1939), a predecessor to the method of Farcasiu (*15*), was used at Ludwigshafen and gave reproducible results. Suida and Motz's method (*14*) involved extraction with pyridine and carbon disulfide (1:1) in a Soxhlet-like apparatus, adsorption onto Terrana (an earth), and elution of the asphaltenes with pyridine after the resins were removed with chloroform. These early reports cautioned that "asphalt resolution methods produce results which are only valid for these methods, i.e. that separation into fractions possesses only relative value." In the late 1930s it was reported at Ludwigshafen that "the resolved fractions are not definite chemical individuals, but groups of substances which depend on the solvent used." These findings are similar to those recently reported by Steffgen et al. (*7*).

Suida and Motz's method was used at Ludwigshafen to compare two coal hydrogenation processes under a variety of conditions. The Welheim Treatment was a coal hydrogenation that did not use gaseous hydrogen, but only the hydrogen donated from a middle distillate oil used as vehicle for the coal. The results of the Welheim Treatment indicated that at 100–150 atm and 350° C Ruhr coal gave 50% asphaltenes (called asphalt). The Ludwigshafen method used gaseous hydrogen at 300 atm and 700 atm and 350° C to give 26% and 7% asphaltenes, respectively, from Upper Silesium coal. Comparisons of various coal conversion processes with a single method have been used most recently by Schwager and Yen (*16*).

Asphaltenes and the Mechanism of Liquefaction

The Germans used the relative distribution of the constituents in the coal hydrogenation product as a measure of the conversion of coal to liquid products. They referred to the variation of the distribution of the coal liquefaction constituents as the degree of hydrogenation (Aufhydrierungsbedingungen).

Rank (*17*) in his studies at Ludwigshafen described rapidly heating bituminous coal to form new substances. "It is thought that these substances precipitable with benzene are primarily products of the decomposition of coal bitumina, in hydrogenation, as well as carbonization, which become soluble in benzene only by further hydrogenation." These findings pre-date by some thirty years the recognition of preasphaltenes (*18, 19, 20*), a class of substances that are formed rapidly from bituminous coal with little or no addition of hydrogen from the gas phase or donor solvent. Preasphaltenes are defined (*20*) as pyridine-soluble, benzene-insoluble material.

With the renewed interest in coal liquefaction, many groups have again taken up research on asphaltenes, following in the spirit of Boussingault's original studies on asphalt. That is, asphaltenes are used as one of the common starting points upon which to build a scientific explanation of the physical and chemical properties of coal conversion products. Although coal-derived asphaltenes are at least as complex as those from asphalts and petroleum crudes, they still convey to the chemist and engineer, by their chemical composition, weight percent, molecular weight, functional group distribution and molecular size, the extent of change taking place during coal hydrogenation. To understand the mechanism of coal liquefaction, it has become increasingly more important to know the details of the chemical structure of asphaltenes. We believe the complexity of this operationally defined fraction cannot be represented by a single model or representative structure. Such average structures have value only when used as tools for developing a working hypothesis. They are composed of an agglomeration of structural features, and should not be taken as literal representations. Asphaltenes may best be described in a phenomenological sense. They reflect their origin, the starting coal, and the process that generated them all at the same time.

Liquefaction Conditions and Asphaltene Character

Many studies have depicted coal liquefaction in terms of the sequential progression through the series of intermediate steps from preasphaltene to asphaltene to oil. Other approaches have emphasized the changes in functional groups over the course of liquefaction. Our recent work (*20, 21, 23, 24, 28*) has concentrated on the fate and changes in the asphaltene fraction of liquid products as the severity of hydrotreatment was varied. In the course of this work, we have developed certain viewpoints on the importance of the character of asphaltenes to physical properties of coal liquids and the relationship of processing history to asphaltene character. What follows is an overview of what we feel are the more important observations.

To illustrate these points, we take up in more detail the relationship between the chemical processing history of coal-derived liquids and the properties of the asphaltenes. Variation of their properties is related in turn to the viscosity of the liquids containing the asphaltenes.

Liquefaction process conditions greatly influence the properties of coal-derived asphaltenes (*21*). When a liquid product containing 31.7% asphaltenes was hydrotreated in a stirred autoclave under a variety of conditions, both with and without a cobalt–molybdenum catalyst, various amounts of conversion to pentane-soluble oils were obtained. It was found that the asphaltenes remaining after hydrotreatment were more aromatic, contained fewer polar functional groups and were of somewhat smaller average molecular size than those present before hydrotreating. These changes may reflect a combination of the selective conversion of relatively hydrogen-rich asphaltene components to oil and the formation of more refractory material by retrogressive reactions. The rate of conversion declined as the amount of conversion increased. Based on this observation, processing asphaltenes to extinction would require either a very long residence time or more severe hydrotreating conditions than commonly used in current liquefaction processes.

The chemical character of the unconverted asphaltenes is also a function of processing. Both the H/C and O/C atomic ratios declined in a regular manner as conversion progressed. In some cases, the oxygen content was reduced to that typically found for pentane-soluble oils (about 2%). At the same time, the atomic H/C ratio of the residual asphaltenes was reduced to values considerably lower (<0.7) than that of the coal. The relative oxygen content may be used as a crude indicator of the severity of hydroprocessing experienced by a particular asphaltene.

Oxygen Functionalities

The importance of oxygen functionalities to the mechanism of conversion of asphaltene may be analogous to that found for coal. Szladow and Given (*22*) have demonstrated that a correlation exists between oxygen removal and conversion of coal to pyridine-soluble material. As shown in Figure 2, there is a correlation between the oxygen content of the asphaltenes and the asphaltene content of the coal-derived liquids. The ether and phenolic oxygen in the asphaltenes declined at about the same rate, at least in the case of this one study (*21*). Whether this is generally true is not known. As a consequence of their parallel rates of decay, the ratio of phenolic oxygen to ether oxygen becomes greater as asphaltene content declines. Since the asphaltenes with lower oxygen content are more resistant to conversion by hydrotreatment, we speculate that oxygen functionalities play an important role in the conversion of asphaltene to oil. The exact manner in which they are involved is still not clear. To clarify the situation, it would help to know if ethers are first

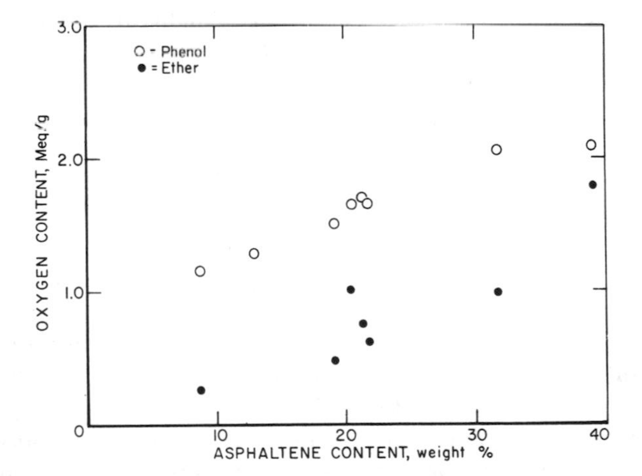

Figure 2. Phenol and ether content vs. asphaltene content for liquid products made under a variety of conditions employing $CoMo/Al_2O_3$ catalyst; Asphaltene contents taken from Ref. 21; phenol content measured by direct titration and ether content taken by difference

converted to phenols or alcohols before elimination of oxygen, or both forms of oxygen are eliminated directly, or the parallel rate of ether and phenol elimination is attributable to the conversion of bifunctional molecules to oil.

Asphaltenes and Viscosity

Asphaltene content bears directly on the physical properties of the liquid product. Viscosity is of particular interest because of the importance of this parameter to operation of liquefaction plants and as a measure of the extent of liquefaction. The correlation between asphaltene content and the viscosity of the liquid has been a subject of a number of investigations (*23–27*). The logarithm of the viscosity ratio, ln η/η_0 (where η and η_0 are the viscosities of the solution and solvent, respectively) was found to be a linear function of concentration when asphaltene was redissolved in the pentane-soluble oil isolated from a coal-derived liquid (*24*). The slopes of these lines, termed the logarithmic viscosity numbers, are a measure of the contribution to the viscosity of a solution attributable to asphaltene. By comparison of logarithmic viscosity numbers of asphaltenes and their acidic and basic subfractions, it was determined that intermolecular association, which is especially strong between the acid and base subfractions, is responsible for a significant portion of the viscosity of these solutions.

Since the logarithmic viscosity number is related to the degree of intermolecular interaction between the polar functional groups of the asphaltene, it is reasonable to expect this parameter to vary with the extent of hydroprocessing given to the liquid product. This point was investigated by measuring the viscosities of solutions of asphaltenes and subfractions of asphaltenes isolated from liquid products that had seen a variety of processing conditions. A standard model solvent system (1-methylnaphthalene/*o*-cresol-88/12) was used to facilitate comparisons. The logarithmic viscosity numbers of eleven samples varied over the range of 3.8–6.8 mL/g. The number-average molecular weights, determined by VPO ranged from 598 to 786. The phenol contents, determined by direct potentiometric titration in pyridine, ranged from 1.00 to 2.12 meq/g. The best fit using both of these parameters in a multiple linear least squares regression analysis is given by the equation:

$$V = 1.86P + 0.0031M - 0.19$$

where V = logarithmic viscosity number, P = phenol content in meq/g, and M = number average molecular weight. Inclusion of both parameters results in a significantly better fit than does either used separately.

Comparison of extrapolations obtained by this equation provides some insight into the relative importance of phenol content and molecular weight to viscosity. For example, a typical asphaltene from a noncatalytic process might have a phenol content of 2 meq/g and a molecular weight of 700. The calculated logarithmic viscosity number is 5.7 mL/g. By extrapolation in the hypothetical case of reducing its molecular weight to 220 (typical for oils) while leaving the phenol content constant, we obtain $V = 4.2$ mL/g. However, by reducing P to zero while leaving M constant, we obtain $V = 2.0$ mL/g. On this basis, one can imagine that selective removal of phenols is potentially a more hydrogen efficient means to reduce the viscosity of a particular liquid than extensive hydrocracking. These considerations may assist in guiding the development of catalysts for the reduction and control of the viscosity of coal-derived liquids.

This summary helps to relate our current outlook to the historical prospective outlined earlier. As a whole, deeper study of the chemical character of asphaltenes requires that the concepts of Boussingault advanced 142 years ago be somewhat refined. The earlier work illustrated the large effects that result simply from variation in asphaltene content. The recent work illustrates that not only the asphaltene content but its chemical structure as well are important to the physical properties of the liquids in which they are found.

Acknowledgment

We gratefully acknowledge the contributions of R. B. LaCount and R. P. Noceti to the study of the viscosity of coal-derived liquids, and S. Friedman for helpful discussions.

Literature Cited

1. Boussingault, J. B. *Ann. Chim. Phys.* **1837,** *64,* 141.
2. Boduszynski, M.; Chadha, B. R.; Pineles, H. *Fuel* **1977,** *56,* 145.
3. Helm, R. V.; Latham, D. R.; Ferrin, C. R.; Ball, J. S. *Chem. Eng. Data Ser.* **1957,** *2*(1), 95.
4. Middleton, W. R. *Anal. Chem.* **1967,** *39,* 1839.
5. Schweighardt, F. K; Thames, B. M. *Anal. Chem.* **1978,** *50,* 1381.
6. Mima, M. J.; Schultz, H.; McKinstry, W. E. "Method for the Determination of Benzene Insolubles, Asphaltenes, and Oils in Coal-Derived Liquids," PERC/RI-76/6, NTIS, U. S. Department of Commerce: Springfield, VA, 1976.
7. Steffgen, F. W.; Schroeder, K. T.; Bockrath, B. C. *Anal. Chem.* **1979,** *51,* 1164.
8. Schultz, H.; Mima, M. *Am. Chem. Soc., Div. Fuel Chem., Prepr.* (Anaheim, CA, Mar., 1978) *23*(2), 76.
9. Puzinauskas, V. P.; Corbett, L. W. "Research Report 78-1," The Asphalt Insitute: College Park, MD, 1978.
10. Marcusson, J. Z. *Angew. Chem.* **1919,** *32,* 113.
11. Berthelot, M. *Bull. Soc. Chim.* **1869,** *11,* 278.
12. Nellensteyn, F. J. *World Pet. Congr., London* **1933,** *2,* 616.
13. Poell, H. *Erdoel Teer* **1931,** 350.
14. Suida, H.; Motz, P. *Petroleum (Berlin)* **1939,** 525.
15. Farcasiu, M. *Fuel* **1977,** *56,* 9.
16. Schwager, J.; Yen, T. F. *Fuel* **1978,** *57,* 100.
17. Rank "The Limits of Solids and Asphalt with Low H_2-Supply in Liquid Phase," in *Tech. Oil Mission Reports* **47,** T-346.
18. Neavel, R. C. *Fuel* **1976,** *55,* 237.
19. Whitehurst, D. D.; Farcasiu, M.; Mitchel, T. O. "The Nature and Origin of Asphaltenes in Processed Coals – Report AF-252"; Electric Power Research Institute: Palo Alto, CA, 1976.

20. Sternberg, H. W.; Raymond, R.; Schweighardt, F. K. *Am. Chem. Soc., Div. Pet. Chem., Prepr.* (New York, Apr., 1976) *21*(1), 198.
21. Bockrath, B. C.; Noceti, R. P. *Fuel Process. Technol.* **1979,** *2,* 143.
22. Szladow, A. J.; Given, P. H. *Am. Chem. Soc., Div. Fuel Chem., Prepr.* (Miami Beach, Sept., 1978) *23*(4), 161.
23. Sternberg, H. W.; Raymond, R.; Akhtar, S. *ACS Symp. Ser.* **1975,** *20,* 111.
24. Bockrath, B. C.; LaCount, R. B.; Noceti, R. P. *Fuel Process. Technol.* **1977/78,** *1,* 217.
25. Thomas, M. G.; Granoff, B. *Fuel* **1978,** *57,* 122.
26. Krzyzanowska, T.; Marzec, A. *Fuel* **1978,** *57,* 804.
27. Taylor S. R.; Li, N. C. *Fuel* **1978,** *57,* 117.
28. Bockrath, B. C.; LaCount, R. B.; Noceti, R. P. *Fuel* **1980,** *59,* 621.

RECEIVED June 23, 1980.

4

Structural Differences Between Asphaltenes Isolated from Petroleum and from Coal Liquid

TEH FU YEN

School of Engineering, University of Southern California, Los Angeles, CA 90007

The structure of asphaltenes derived from petroleum is different from those of coal. In general, coal-derived asphaltenes have higher aromaticity than petroleum-based asphaltenes. The aromatic system within the petroleum asphaltene is largely peri-condensed, while that of the coal asphaltene is kata-condensed. The substitution of the peripheral carbons in the aromatic system of coal-derived asphaltenes is less than and shorter than that in asphaltenes of petroleum origin. The unit molecular weight of coal asphaltenes is 400–600, whereas that of petroleum asphaltenes is 800–2500, which is reflected in stacking height as well as layer diameter. Another major characteristic difference is that coal asphaltenes contain more hydroxyl and pyrrolic groups in addition to ether–oxygen or basic nitrogen functions than do petroleum asphaltenes.

Asphaltene is an essential component of any dark-colored, heavy, viscous and nonvolatile oil, regardless of the oil source. Asphaltene can be obtained from the oil extracted from a naturally occurring organic-rich fossil material by a simple solvent fractionation. Asphaltene also can be obtained from the chemical conversion product of a solid fuel, such as pyrolysis or catalytic hydrogenation of coal or shale. The former is an example of the asphaltene isolated from native petroleum oil. An example of the latter is the asphaltene obtained from a synthetic crude, such as shale oil or coal liquid.

In this chapter, I emphasize the origin or source from which the asphaltenes came, rather than the environment or conditions to which they were exposed. For example, I will group asphaltenes from refinery bottoms with petroleum-derived asphaltenes in the discussion. Similarly, asphaltenes from the solvent extracts of raw coal are also classified as coal-derived asphaltenes.

0065-2393/81/0195-0039$05.00/0

Concept of Average Structure

For well-defined, simple organic molecules, the chemical structure can be elucidated by composition analysis and some spectroscopic methods. As the molecule becomes more complex in nature, the traditional chemical structures do not apply. New methodology must be developed to characterize them. The following is a list of some general classes of substances ranging from simple to complex:

1. *simple compounds*—organic molecules described in most general textbooks or handbooks, along with alkaloids, vitamins, and drugs.
2. *regular polymers*—substances containing fixed repeating sequences, for example, polystyrene.
3. *random polymers*—substances similar to those in 2, with the exception that their building blocks occur in random fashion, such as lignin and melanin.
4. *intrinsic mixtures*—generally, mixtures of similar molecules, perhaps isomers or homologs, of a narrow carbon-number distribution range associated by intermolecular force; petroporphyrins and coal tar pitch would be good examples.
5. *multipolymers*—the most complex substances known, which are referred to as polycondensates with different repeating blocks, with copolymers containing 2 blocks, terpolymers 3, and multipolymers expected to contain *n* blocks. Because of the inter- and intramolecular interactions within the structure, this class also has the nature of class 4, with association between molecules. Asphaltenes generally belong to this class.

Because of the variance in multipolymers, an exact chemical structure is not possible. To differentiate between different asphaltenes, the methodology leading to an average structure is necessary.

Microstructure Versus Macrostructure

For a discrete molecule with a simple structure, a microstructure is sufficient to characterize the given molecule. For a complex system such as that of asphaltene, the information required for characterization has to include association as well as micelle formation. The microstructure has been chosen arbitrarily to refer to short-range bonding, that is, distances between 0.5 Å–2.0 Å; whereas the macrostructure (bulk structure) pertains to molecular interactions or orders at larger distances (20 Å–2000 Å).

Structural Parameters

To characterize the most complex system, many selected physical and chemical methods should be employed. Simply, the reason is that one given method may provide only partial structural information about the system.

What is important is that, if different methods (with or without certain assumptions) yield identical results, multiple methods will increase the precision of structural information.

Structural information obtained from various methods can usually be represented by a set of structural parameters, S_i. These parameters are related to the physical and chemical properties of a given substance. A given set of properties, P_j, is unique to a given substance. Therefore, a given set of structural parameters can be used to characterize a given asphaltene:

$$(S_i) = f_1(P_j)$$

Both structural parameters and properties derived in this manner are in matrix form. Furthermore, the production or genesis of asphaltene must depend on a set of environmental or refinery variables, such as temperature, and pressure. These quantities, tentatively called refinery variables, R_k, can affect the nature of the chemical and physical properties of asphaltenes:

$$(R_k) = f_2(P_j)$$

Consequently, by studying the structural parameters of asphaltenes, one can directly relate them to refinery variables. The application to engineering problems is that by controlling a given set of refinery variables, the properties of the asphaltenes can be controlled.

In certain cases a structural parameter can be obtained by several methods. For example, aromaticity, f_a, can be obtained from x-ray, proton NMR, and ^{13}C NMR. The precision of f_a is governed by the differences in upper and lower limits of the three methods. To test the validity of a structural parameter, known compounds or model systems should be applied to such tests as controls, with the simpler compounds supplied first. (As an illustration, the parameter VPO molecular weight has been determined for three different solvents and extrapolated to infinite dilution. Also, the data points have been verified by using a model for computation) (Figure 1) (*1*).

Petroleum-Derived Asphaltenes

Microstructure. By using a group of structural parameters obtained through a number of physical methods (such as x-ray, NMR, MS, IR, VPO, DTA, densimetric methods, EM, and SAS) and chemical methods (such as oxidation, alkylation, and halogenation), the structure of petroleum asphaltene has been gradually revealed. As an example, the structural parameters of an asphaltene derived from a Laquinillas crude oil from Venezuela (API gravity 20°, Conradson carbon number 13.39, Miocene Age) are listed in Table I. An empirical formula can be deduced.

The value of ring condensation index (*3*), ring compactness (*7*), and the extent of condensation of unsubstituted aromatic clusters (*4*) all indicate a skeleton of a peri-condensed system in the aromatic portion of this asphaltene.

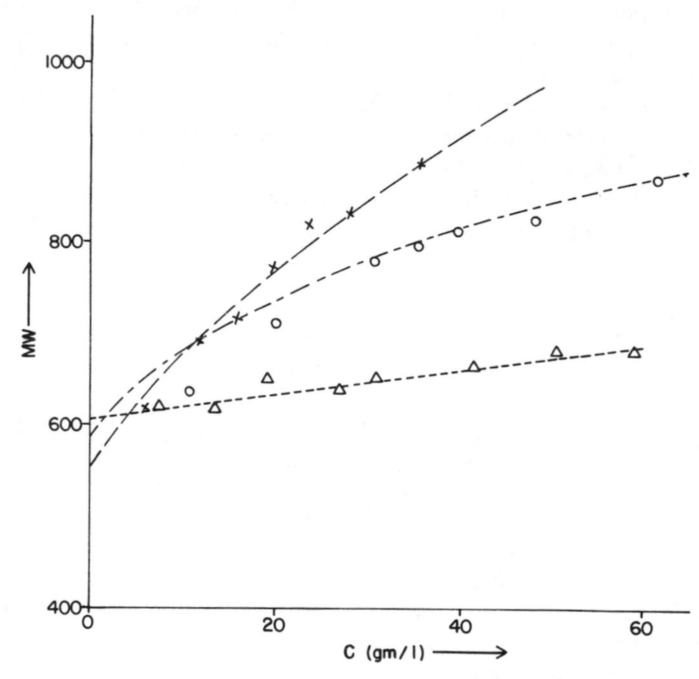

Figure 1. Molecular weight determination of Synthoil coal liquid asphaltene at different concentrations in three solvents: in benzene (x) experimental, (— —) calculated; in chloroform (○) experimental, (— - —) calculated; in tetrahydrofuran (Δ) experimental, (- - -) calculated

This work is supported by oxidation rate studies (*9*), which found similar rates for carbons and graphite. The aromaticity value of this asphaltene (*5*) indicated that 60% of the carbons are found in the saturated form. From IR studies, the amount of total methyl groups and (from conjunctive NMR work) (*4*) the distribution of the short-chain to the aromatic portion were obtained. The value of the condensation of naphthenic clusters, obtained by NMR (*4*), reveals that the system is situated between peri- and kata-condensation. The average chain length, as determined by NMR and, especially, through IR studies (*10*), provides an indication of a long-chain paraffinic substituent. From ESR studies (*11*), it was found that most heteroatoms are associated with the aromatics, although some sulfur can be found in the saturated portion. Acidic protons were not observed to be associated with heteroatoms. A hypothetical structure is presented in Figure 2 for the sake of illustration and economy of description. The exact location of the groups represented in the illustration are relative.

Macrostructure. Generally, petroleum-derived asphaltene is unisotropic in structure. It can be classified as a mesomorphic liquid, that is, many individually oriented clusters are suspended and randomly distributed within their lower molecular weight homologs (*12*). The individual clusters are

Table I. Examples of the Structural Parameters for the Microstructure of a Petroleum-Derived Asphaltene

Structural Parameters	*Methods*	*Reference*	*Experimental*	*Calculated*
M	VPO	(2)	4750	4276
%C	elemental analysis	(3)	84.2	83.0
%H	elemental analysis	(3)	7.9	8.1
%O	direct determination	(3)	1.6	1.5
%N	elemental analysis	(3)	2.0	1.3
%S	elemental analysis	(3)	4.5	6.0
X/C	elemental analysis	(11)	0.055	0.054
$H_{ar}^{\circ}(H_A/H, h_a)$	NMR	(4)	0.055	0.046
$H_{\alpha}^{\circ}(H_{\alpha}/H)$	NMR	(4)	0.27	0.25
$h_s(H_s/H)$	NMR	(4)	0.95	0.95
H_R/H	NMR	(4)	0.31	0.31
H_{sme}/H	NMR	(4)	0.19	0.21
$H_N/H\ (h_n)$	NMR	(4)	0.18	0.18
f_a	x-ray	(5)	0.41	0.42
L_a	x-ray	(5)	9.7 Å	9.85 Å
$H_{aru}/C_{ar}\ (H_I/C_A)$	NMR	(4)	0.43	0.44
$\sigma(C_{su}/H_I)$	NMR	(4)	0.71	0.71
$n\ (C_S/C_{su})$	NMR	(4)	3.2	4.3
H_N/C_N	NMR	(4)	1.25	1.06
C_{sme}/C_p	NMR	(4)	0.47	0.50
$C_p/C\ (f_p)$	IR	(6)	0.35	0.36
$C_N/C\ (f_n)$	IR	(6)	0.24	0.22
C_{Me}/C	IR	(6)	0.15	0.16
C_{mp}/C_{su}	IR	(6)	1.7	1.1
C/R	densimetric	(3)	6.2	5.0
F/C_A	densimetric	(7)	0.32	0.60
Λ_m	MS	(2)	576	524
N	MW and x-ray	(8)	4	4
Empirical formula		(8)	Laquinillas asphaltene	$(C_{74}H_{87}NS_2O)_4$

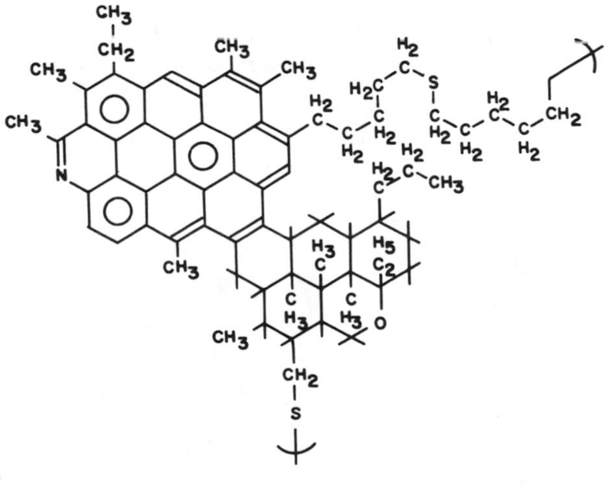

Figure 2. Hypothetic structure for a Laquinillas asphaltene

ordered and consist of a number of planar aromatic molecules stacked in layers via the π–π association. In general, the stacking occurs with five or six layers, as discovered by x-ray diffraction. Some commonly used structural parameters for the macrostructure of petroleum-derived asphaltene are shown in Table II.

The conventional meaning of molecular weight does not apply to the case of petroleum asphaltene; however, depending on the methods used, a unit sheet weight (1000–4000), a cluster or particle weight (4000–10,000), and a micelle weight (40,000–40,000,000) are possible. Most intramolecular associated particles can be taken as the basic molecular weights, such as in the case of the uniparticle evaporation (*13*). Between the sheets of the stacks, small planar aromatic molecules can be inserted. Also, interchelation complexes can be made (*14*) in such a way that many catalytic effects may be explained (*15*). The association–disassociation energy of the sheets is 14–20 kcal/mol (*16*). These interactions involve charge-transfer mechanisms, as found by studies using both IR spectroscopy (*17*) and ESR (*18*).

Many of the aromatic portions of the sheet contain defective centers (gaps and holes) (*19*), and usually these are the coordination centers for metals (*20*). Schematically, the arrangement of sheets with the zig-zag chains can be best illustrated as in Figure 3. The π–π association will result in the production of a crystallite (Figure 3 **A**). The zig-zag chain can orient in bundle-like fashion (**B**) depending on the torsional angle when the propagation of sp^3-bonded orbital takes place. These island-like stacks are termed particles (**C**). Usually, particles can be further associated to micelles (**D**); the peripheral groups are polar. There are a number of weak links (**E**) along the chain configuration; some of these are ether, thioether, or even hydrogen bonds, which can undergo cleavage. There are imperfections in the aromatic systems with di- or trivalent heteroatoms, such as nitrogen, sulfur, or oxygen. Clusters may also be linked

Table II. Some Commonly Used Structural Parameters for the Macrostructure of a Petroleum-Derived Asphaltene

Structural Parameters	*Methods*	*Reference*	*Value*
L_c	x-ray	(*5*)	16–20 Å
M_e	x-ray	(*5*)	5–6
D_A	EM	(*13*)	19–37 Å
D_G	EM	(*13*)	7–97 Å
R	SAS	(*21*)	33
D	SAS	(*21*)	86
$\underline{l}_c$	SAS	(*21*)	300–400 Å
$\bar{l}$	SAS	(*21*)	500–600 Å
r_M	ultracentrifuge	(*2*)	22 Å
D_e	low angle scattering	(*21*)	50 Å
M_f	film balance	(*22*)	80,000–140,000

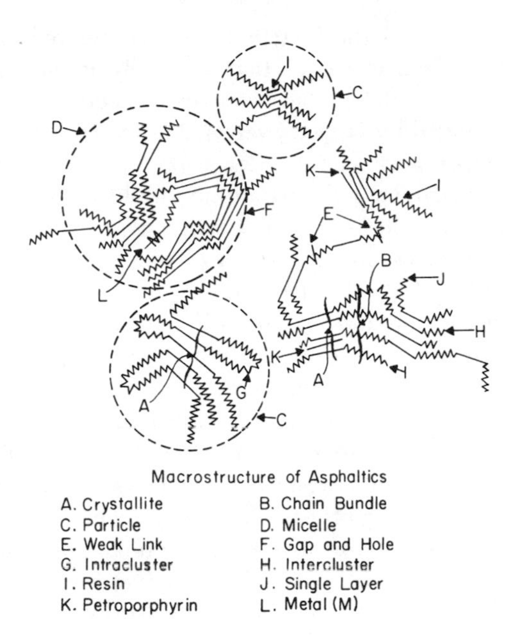

Figure 3. The general features of the macrostructure of petroleum-derived asphaltenes and related substances

by folding of the chains (intracluster **G**) or by bridging through weak links (intercluster **H**). The resin (**I**) is a small homolog of asphaltene that is larger, and the molecular weight is higher. A simple layer (**J**) of the stacked clusters can be formed by disassociation, provided that conditions do not favor the reassociation. Petroporphyrins (**K**) are a single layer containing extended conjugation of pyrroles. The metal (**L**) can be a coordination center for several clusters.

Coal-Derived Asphaltene

Microstructure. The characterization of coal-derived asphaltene is quite similar to that of petroleum-derived asphaltene. Since it is anticipated that coal-derived asphaltene will have acid/neutral and base characteristics (*26*, *38*), the average structure of both must be considered. In Table III, Structure I is amphoteric (or slightly basic), and Structure II is an acid/neutral representation. A mixture of both may be typical of the average structure of a coal-derived asphaltene. At present, we will illustrate this by an asphaltene obtained from coal liquid of the Synthoil process. (The coal is hvAb, West Kentucky, Homestead Seam; the coal liquid is obtained by catalytic hydrogenation at 450° C and 4000 psig having %C, 86.7; %H, 8.38; %N, 0.93; %S, 0.09; %O, 3.2; and %Ash, 0.7.)

Especially important is the fact that the coal-derived asphaltenes definitely have a lower molecular weight, that is, usually on the order of 500–600. They also contain a much smaller aromatic system that appears to be kata-condensed as judged by H_{aru}/C_{ar} value (*24*), the ring condensation index (*30*), and ring compactness (*30*). Investigations by ^{13}C NMR (*24*) and IR spectroscopy (*36*) reveal no long paraffinic substituents. In general, coal-derived asphaltenes have 30%–50% substitution at the peripheral position of the aromatic system. Consequently, the asphaltene is expected to be more reactive. Both the phenolic and pyrrolic groups are abundant in coal-derived asphaltene (*32*). Structures I and II, utilized for the calculation of the last two columns of Table III, are shown in Figure 4. These structures are consistent with those reported in recent literature (*46*).

Macrostructure. A very distinctive feature of coal-derived asphaltenes is that they appear to be associated both in concentrated and diluted solutions, and, particularly, in nonpolar solvents. This fact is supported by the viscosity measurement (*40*). The fitting of a model of monomer–dimer–trimer for the coal-derived asphaltenes is almost perfect (*34*). The hydrogen-bonding nature of the interaction within the coal-derived asphaltenes is important (*35, 41, 42*). The other interaction is still the π–π association. X-ray data give a low value of

Table III. Examples of the Microstructure of a Coal-Derived Asphaltene

Structural Parameters	*Methods*	*Reference*	*Experimental*	*Calculated I*	*Calculated II*
M	VPO	(*25*)	561	549	559
%C	elemental analysis	(*23*)	87.27	85.26	88.80
%H	elemental analysis	(*23*)	6.51	6.38	6.62
%N	elemental analysis	(*23*)	1.63	2.40	2.50
%S	elemental analysis	(*23*)	0.66	0	0
%O	elemental analysis	(*23*)	3.93	5.83	2.86
X/C	elemental analysis	(*11*)	0.044	0.077	0.048
H°_{ar} (H_A/H, h_a)	NMR	(*24*)	0.31	0.29	0.30
H°_{α} (H_{α}/H)	NMR	(*24*)	0.42	0.40	0.43
H°_{O}	NMR	(*24*)	0.25	0.29	0.27
f_a	^{13}C NMR	(*24*)	0.69	0.69	0.68
H_{aru}/C_{ar} (H_I/C_A)	NMR	(*24*)	0.67	0.71	0.66
σ (C_{su}/H_I)	NMR	(*24*)	0.45	0.47	0.47
n (C_S/C_{su})	NMR	(*24*)	1.6	1.5	1.6
α – Me	^{13}C NMR	(*30*)	1.91	2	2
β – Me	^{13}C NMR	(*30*)	1.3	1	1
N	^{13}C NMR	(*30*)	1.9	2	2
C/R	densimetric	(*30*)	6.57	6.5	5.9
F/C_A	densimetric	(*30*)	–0.08	0	0.036
M	GPC	(*31*)	548	549	559
L_a	x-ray	(*28*)	10.3 Å	7.0 Å	7.4 Å
$\%O_{O-H}$	IR	(*32*)	2.67	2.91	2.86
$\%N_{N-H}$	IR	(*32*)	0.93	0	2.50
C_{au}	refractive index	(*39*)	14	14	13
Empirical formula		(*29*)	Synthoil asphaltene	$C_{39}H_{35}NO_2$	$C_{41}H_{37}ON$

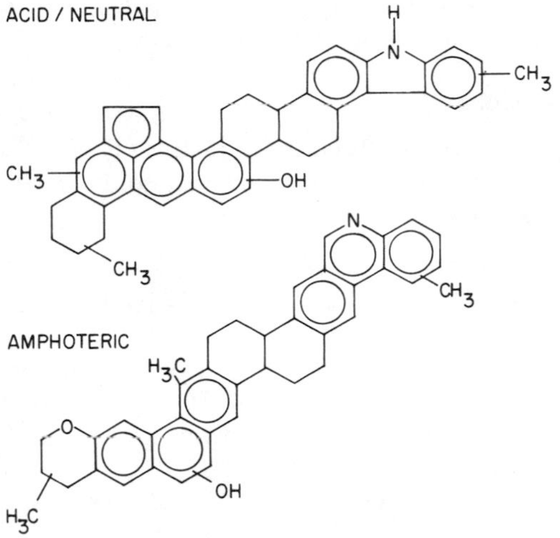

Figure 4. Hypothetical structures for the Synthoil asphaltenes of both acid/neutral and amphoteric fractions

three to four layers, probably because of the small groups of aromatic rings (*29*, *34*). The support for the π–π association is best exemplified by the charge-transfer properties of I_2 and TCNE (*33*), along with their ESR properties. Structural parameters for the characterization of the macrostructure of coal-derived asphaltenes are shown in Table IV.

The weak linkage in coal and coal-derived products may be attributable to excessive inter- and intramolecular bonding (*41*). The fact that the solubilization and swelling curves are similar is important (*43*). This type of bonding may increase its extent of crosslinking so much that carboid (or preasphaltene) is no longer benzene-soluble. So far, there is no exciton interaction for an exponential temperature dependence of the spins for ESR work (*37*, *44*). There has been limited work on the long-range order of petroleum-derived asphaltene. By using the shadowing technique of TEM, the height of the synthoil carboid was found to be 10–30 Å. This order of discontinuity is not

Table IV. Parameters for the Characterization of the Macrostructure of Coal-Derived Asphaltenes and Related Derivatives

Structural Parameter	*Methods*	*Reference*	*Value*
L_c	x-ray	(*29*)	10.5–14 Å
M_e	x-ray	(*29*)	4
M_i	charge-transfer	(*33*)	5/2
M_a	association	(*34*)	3
h	TEM	(*39*)	10–30 Å
W	TEM	(*39*)	100–200 Å

unreasonable, in view of the 20-Å band in coal and in the pyrine-soluble fraction of vitrinite as an intercrystallite distance (*45*).

Summary and Conclusion

The particular example discussed above typifies the asphaltenes so far studied. The following general statements can be deduced from the average structure so far obtained (*47*).

1. The aromaticity of petroleum-derived asphaltene (f_a = 0.2–0.5) is lower than that of coal-derived asphaltene (f_a = 0.6–0.7).
2. The aromatic ring systems within petroleum-derived asphaltene are much more condensed (H_{aru}/C_{ar} = 0.3–0.5, which is peri) than those of coal-derived asphaltene (H_{aru}/C_{ar} = 0.5–0.7, which is kata).
3. The substituents of the petroleum-derived asphaltenes are longer (n = 4–6) than those of coal-derived asphaltenes (n = 1–2).
4. The aromatic system of petroleum-derived asphaltene is extensively substituted (50%–70%) whereas the coal-derived asphaltene is sparingly substituted (30%–50%).
5. The molecular weight of petroleum-derived asphaltene is about ten times higher than that of the coal-derived asphaltene. Unit molecular weight of coal asphaltene is 400–600 whereas that of petroleum asphaltene is 800–2500.
6. Petroleum-derived asphaltene is less reactive to physical or chemical agents than is coal-derived asphaltene.
7. Petroleum-derived asphaltene is more highly associated (Me = 5–7) than is coal-derived asphaltene (Me = 2–4); this will be reflected in the ease of processing.
8. The aromatic system in coal-derived asphaltene is small (L_a = 7–14 Å), compared with that of petroleum (L_a = 10–15 Å).
9. Petroleum-derived asphaltene is less polar than coal-derived asphaltene (X/C for coal-derived asphaltene is about 0.08, and for petroleum-derived asphaltene is 0.05).
10. Coal-derived asphaltene contains more hydroxyl and pyrrolic groups in addition to ether-oxygen or basic nitrogen functions.
11. The high polarity and low association of coal-derived asphaltenes can be used to explain the nature (hydrogen-bonding) and reactivity of coal conversion.
12. The charge-transfer nature of donor–acceptor properties of petroleum asphaltenes can influence the processing of petroleum.

Acknowledgment

Support from the U.S. Department of Energy, under contracts EX-76-01-2031, EF-77-G-01-2738, and ET-78-G-01-3379, and from Gas Research Institute, under contract 5010-362-0036, is acknowledged.

Glossary

- A_m Aromatic Sheet Weight from Mass Spectrometry
- C Number of Carbons
- C/R Ring Condensation Index
- C_A Number of Aromatic Carbons
- C_{ar} $C_{ar} = C_A$, another converter
- C_{au} Aromatic Carbons Per Structural Unit
- C_{me} Number of Methyl Carbons
- C_{me}/C Methyl Content
- C_{mp} Number of Methylenic Carbons
- C_{mp}/C_{su} Chain Length
- C_N Number of Naphthenic Carbons
- C_P Number of Paraffinic Carbons
- C_s Number of Saturated Carbons ($C_s = C_p + C_N = C - C_A$)
- C_{sme} Number of Saturated Methyl Carbons
- C_{su} Number of Substitutable Carbons in Aromatic System
- D Micelle or Particle Diameter
- D_A Arithmetic Mean Diameter of Particles
- D_e Low Angle Scattering Distance
- D_G Geometric Mean Diameter of Particles
- F Number Interval Quaternary Carbon in Aromatic System
- F/C_A Degree of Fusion
- f_a Aromaticity ($f_a = C_A/C$)
- f_n Naphthenicity
- f_p Paraffinity
- f_s Fraction of Saturated Carbon to Total Carbon ($f_s - f_n + f_p - 1 - f_a$)
- h Cluster Height
- H Number of Hydrogen
- h_a Hydrogen Aromaticity ($h_a = H_A/H$)
- H_A Number of Aromatic Hydrogens
- H_α Number of α-Substituted Hydrogens Next to an Aromatic System
- H_α° $H_\alpha^\circ = H_\alpha/H$
- H_{ar}° $H_{ar}^\circ = H_A/H = h_a$
- H_{aru} Number of Fully Substitutable Hydrogens in the Aromatic System
- H_I $H_I = H_{aru}$
- H_I/C_A Extent of Condensation
- H_N Number of Naphthenic Hydrogens
- H_O° $H_O^\circ = 1 - (H_{ar}^\circ + H_\alpha^\circ)$
- H_p Number of Paraffinic Hydrogens
- H_R Number of Methylenic Hydrogens
- h_s Hydrogen Non-Aromaticity ($h_s = 1 - h_a$)
- H_s Number of Saturated Hydrogens ($H_s = H_N + H_P$)
- $\bar{l}$ Inhomogeneity Length

L_a Layer Diameter of the Aromatic System
l_c Coherence Length
L_c Cluster Diameter
M VPO Molecular Weight
M_a Association Layer
M_e Effective Layer
α-Me Number of α-Substituted Methyls
β-Me Number of β-Substituted Methyls
M_f Molecular Obtained Film Balance
M_i Charge-Transfer Layer
n Average Chain Length ($n = C_s/C_{su}$)
N Number of Nitrogens
$\overline{N}$ Oligomer Number
$\%N_{N\text{-}H}$ Imino Content
O Number of Oxygens
$\%O_{O\text{-}H}$ Phenolic Hydroxyl Content
R Number of Rings
$\overline{R}$ Radius of Gyration
r_M Radius of Particle in Sedimentation
S Number of Sulfurs
W Width of Clusters
X Number of Heteroatoms ($X = S + N + O$)
σ Degree of Substitution ($\sigma = C_{su}/H_I$)
% Weight Percent

Literature Cited

1. Lee, W. C.; Schwager, I.; Yen, T. F. *Anal. Chem.* **1977,** *49,* 2363–2365.
2. Dickie, J. P.; Yen, T. F. *Anal. Chem.* **1967,** *39,* 1847–1852.
3. Erdman, J. G.; Hanson, W. E.; Yen, T. F. *J. Chem. Eng. Data* **1961,** *6,* 443–448.
4. Erdman, J. G.; Yen, T. F. *Am. Chem. Soc., Div. Pet. Chem., Prepr.* (Wash. D.C., Mar., 1962) 7(3), 99–111.
5. Erdman, J. G.; Pollack, S. S.; Yen, T. F. *Anal. Chem.* **1961,** *33,* 1587–1594.
6. Erdman, J. G.; Yen, T. F. *Am. Chem. Soc., Div. Pet. Chem., Prepr.* (Wash. D.C., Mar., 1962) *7*(1), 5–18.
7. Dickie, J. P.; Yen, T. F. *J. Inst. Pet.* **1968,** *54,* 50–53.
8. Yen, T. F. *Energy Sources* **1974,** *1*(4), 447–463.
9. Erdman, J. G.; Ramsey, V. G. *Geochim. Cosmochim. Acta* **1961,** *25,* 175.
10. Yen, T. F. *Nature Phys. Sci.* **1971,** *233*(9), 36.
11. Sprang, S. R.; Yen, T. F. *Geochim. Cosmochim. Acta* **1977,** *41,* 1007–1018.
12. Yen, T. F., presented at the *13th Biennial Conf. Carbon, Irvine, CA 1977,* 322.
13. Dickie, J. P.; Haller, N. M.; Yen, T. F. *J. Coll. Interface Sci.* **1969,** *29,* 475.
14. Sill, G. A.; Yen, T. F. *Fuel* **1969,** *43,* 161.
15. Yen, T. F. *Energy Sources* **1978,** *4*(3), 339.
16. Tynan, E. C.; Yen, T. F. *Fuel* **1969,** *43,* 191.
17. Yen, T. F. *Fuel* **1973,** *52,* 93.
18. Yen, T. F.; Young, D. K. *Carbon* **1973,** *11,* 33.
19. Erdman, J. G.; Saraceno, A. J.; Yen, T. F. *Anal. Chem.* **1962,** *34,* 694.
20. Boucher, L. J.; Dickie, J. P.; Tynan, E. C.; Vaughan, G. B.; Yen, T. F. *J. Inst. Pet.* **1969,** *55,* 87.

21. Pollack, S. S.; Yen, T. F. *Anal. Chem.* **1970,** *42,* 623.
22. Pfeiffer, J. P. H.; Sall, R. N. J. *J. Phys. Chem.* **1940,** *44,* 139.
23. Schwager, I.; Yen, T. F. In "Liquid Fuel from Coal," Ellington, R. T., Ed.; Academic: New York, 1977; pp. 233–248.
24. Farmanian, P. A.; Schwager, I.; Yen, T. F. In "Analytical Chemistry of Tar Sands, Oil Shale, Coal, and Petroleum," *Adv. Chem. Ser.* **1978,** *170,* 66–77.
25. Schwager, I.; Yen, T. F. *Fuel* **1978,** *57,* 100.
26. Schwager, I.; Yen, T. F. *Fuel* **1979,** *58,* 219.
27. Kwan, J. T.; Saade, H.; Schwager, I.; Worral, J.; Yen, T. F., presented at the *174th Nat. Meet., Am. Chem. Soc., Chicago, Aug.–Sept., 1977.*
28. Kwan, J. T.; Yen, T. F. *Am. Chem. Soc., Div. Fuel Chem., Prepr.,* (San Francisco, Aug.–Sept., 1976) *21*(7), 67.
29. Farmanian, P. A.; Kwan, J. T.; Schwager, I.; Yen, T. F., unpublished data.
30. Yen, T. F. "Chemistry and Structure of Coal-Derived Asphaltenes," U.S. DOE, Quarterly Progress Report for Contract No. EX-76-C-01-2031, Apr.–June, 1978.
31. Kwan, J. T.; Lee, W. C.; Meng, S.; Schwager, I.; Yen, T. F. *Anal. Chem.* **1979,** *51,* 1803.
32. Schwager, I.; Yen, T. F. *Anal. Chem.* **1979,** *51,* 569.
33. Kwan, J. T.; Miller, J. F.; Schwager, I.; Yen, T. F. *Am. Chem. Soc., Div. Fuel Chem., Prepr.* (Anaheim, Mar., 1978) *23*(1), 284.
34. Lee, W. C.; Schwager, I.; Yen, T. F. *Am. Chem. Soc., Div. Fuel Chem., Prepr.* (Anaheim, Mar., 1978) *23*(1), 284.
35. Yen, T. F., *Salt. Res. Ind., Workshop on Coal Chemistry, Prepr.* (1976) 144–164.
36. Farmanian, P. A.; Kwan, J. T.; Lee, W. C.; Miller, J. G.; Schwager, I.; Weinberg, V.; Yen, T. F., presented at the *Conf. Anal. Chem. Appl. Spectrosc., Pittsburgh, 1978,* 49.
37. Kwan, C. L.; Yen, T. F. *Anal. Chem.* **1979,** *51,* 1225.
38. Raymond, R.; Schweighardt, F. K.; Sternberg, H. W. *Science* **1975,** *188,* 49.
39. Yen, T. F. "Chemistry and Structure of Coal-Derived Asphaltenes," U.S. DOE, Quarterly Progress Report for Contract No. EX-76-C-01-2031, July–Sept., 1978.
40. Kan, N. S.; Li, N. C.; Susco, D. M.; Tewari, K. C. *Anal. Chem.* **1979,** *51,* 182.
41. Lee, W. C.; Yen, T. F. *Proc. Intersociety Energy Conversion Eng. Conf., 14th, Boston, 1979.*
42. Brown, B. J.; Galya, L. G.; Li, N. C.; Taylor, S. R. *Spectrosc. Lett.* **1976,** *9,* 733.
43. Weinberg, V.; Yen, T. F., presented at the *175th Nat. Meet. Am. Chem. Soc., Anaheim, Mar., 1978.*
44. Friedel, R. A.; Hough, M.; Retocfsky, H. L.; Thompson, C. P. *Am. Chem. Soc., Div. Fuel Chem., Prepr.* (Chicago, Aug.–Sept., 1977) *22*(5), 90.
45. Hirsch, P. B. *Proc. R. Soc.* **1954,** *A226,* 143.
46. Charlesworth, J. M. "Structure of Coal Derived Asphaltenes," *4th Aust. Workshop on Coal Hydrogenation, Richmond, Victoria, 1979,* V13.
47. Yen, T. F. In "The Future of Heavy Crude Oils and Tar Sands"; (Meyer, R. F.; Steele, C. T., Eds.; McGraw-Hill: 1980; pp. 174–179.

RECEIVED September 9, 1980.

5

Consequences of the Mass Spectrometric and Infrared Analysis of Oils and Asphaltenes for the Chemistry of Coal Liquefaction

S. E. SCHEPPELE[1]—Department of Chemistry, Oklahoma State University, Stillwater, OK 74074 and Bartlesville Energy Technology Center, U.S. Department of Energy, Bartlesville, OK 74003

P. A. BENSON, G. J. GREENWOOD, and Q. G. GRINDSTAFF—Department of Chemistry, Oklahoma State University, Stillwater, OK 74074

T. ACZEL and B. F. BEIER—Exxon Research and Engineering Company, Baytown, TX 77520

Oils and asphaltenes are generally assumed to be key intermediates in coal liquefaction. Since fundamental chemical/mathematical principles require compositionally unique fractions, mass spectrometry and infrared spectroscopy were employed to obtain detailed molecular analyses of fractionated coal-liquid-derived oils and asphaltenes. The asphaltenes contain higher molecular weight homologs in many specific–Z series and different compound types than do the oils. However, both fractions contain appreciable quantities of compound types equivalent in molecular formula and, hence, assumably in molecular structure. The chemical/physical properties of oils and asphaltenes are not describable in terms of average (hypothetical) molecular structures. The lack of unique oil and asphaltene compositions necessitates detailed molecular analysis as a prerequisite for understanding, assessing, and controlling the chemical/physical phenomena intrinsic to production and processing of coal liquids.

The concept of preasphaltenes, asphaltenes, and oils is of historical significance in the development of the science and technology of coal liquefaction (*1*). This concept is routinely employed in attempts to understand, assess,

[1]To whom correspondence should be addressed at the Bartlesville Energy Technology Center.

0065-2393/81/0195-0053$07.50/0

and control the chemical/physical phenomena intrinsic to the production and utilization of coal liquids (*1–28*). A necessary prerequisite for this approach to be valid and meaningful is that preasphaltenes, asphaltenes, and oils correspond to chemically well-defined fractions obtained from solvent separation of coal liquids by standard and reproducible procedures. Unfortunately, the paucity of detailed molecular characterization data for these fractions precludes an assessment of the extent to which the compositions of preasphaltenes, asphaltenes, and oils reflect the composition of the coal liquid and chemical/physical phenomena involved in the solvent separation. Consequently, we initiated a detailed molecular group-type analysis of the oils and asphaltenes obtained by simple solvent extraction of a Char-Oil-Energy Development (COED) coal liquid. Kinetic studies of coal liquefaction (*29*), spectroscopic characterization of coal liquids (*30, 31*), dependence of the yields of solubility fractions on extraction sequence (*32*), and analysis of the physical chemistry involved in solvent fractionation (*33*), which were published subsequent to the initiation of this research, question the unique chemical significance of oils, asphaltenes, and preasphaltenes obtained by classical extraction methods (*1, 7, 16, 17, 30, 32, 34*). Consequently, the results obtained in this study assume additional significance.

Experimental

Since the experimental procedures have been described in detail (*35*), they are presented here in summary form.

Separation Procedures. As shown in Figure 1, a sample of a coal liquid derived from Colorado Bear-Mine coal by the COED process (*36*) was separated into 25% oils (pentane solubles), 57% asphaltenes (pentane insolubles/benzene solubles), and 18% residue. The separation of the oils into acids, bases, and neutrals using ion exchange chromatography (*37*) and the removal of saturates from the neutral fraction by silica gel chromatography is illustrated in Figure 2. As seen in Figure 3, the asphaltenes were similarly separated into acids, bases, and neutrals using benzene as the nonaqueous eluent. Removal of the acids from the asphaltenes required three column volumes of anion exchange resin. The saturates present in the 88.2% of the asphaltene neutrals extractable with pentane were removed using silica gel chromatography.

Spectroscopic Methods. Infrared spectra of the various aromatic concentrates were recorded at room temperature on a grating instrument (Beckman IR-7) covering the spectral range 600–4000 cm^{-1}. Solution spectra were obtained using 0.5-mm matched compensating cells and $CDCl_3$ as the solvent, except for the oil and asphaltene neutrals for which CCl_4 was the solvent. Matrix-isolated IR spectra of the oil neutrals, acids, and bases and of the asphaltene neutrals and acid fractions **A** and **C** were acquired (*38*) using KBr windows. The IR spectrum of asphaltene acid fraction B, because of its insolubility, was obtained using a KBr pellet.

Mass spectral data were obtained for the various oil and asphaltene aromatic fractions using a CEC 21-110B mass spectrometer equipped with a modified combination field-ionization/electron-impact (FI/EI) ion source; and for the asphaltene fractions using a KRATOS MS-50 mass spectrometer equipped with a low-voltage electron-impact (LV/EI) ion source. With the former instrument, 70-eV EI/MS were recorded at a static resolution of approximately 21,000 on Ilford Q2 photographic plates, and FI/MS were produced with an emitter potential of 5.8 kV and a counterelectrode potential of +800 V to −800 V. Low-resolution FI/MS were

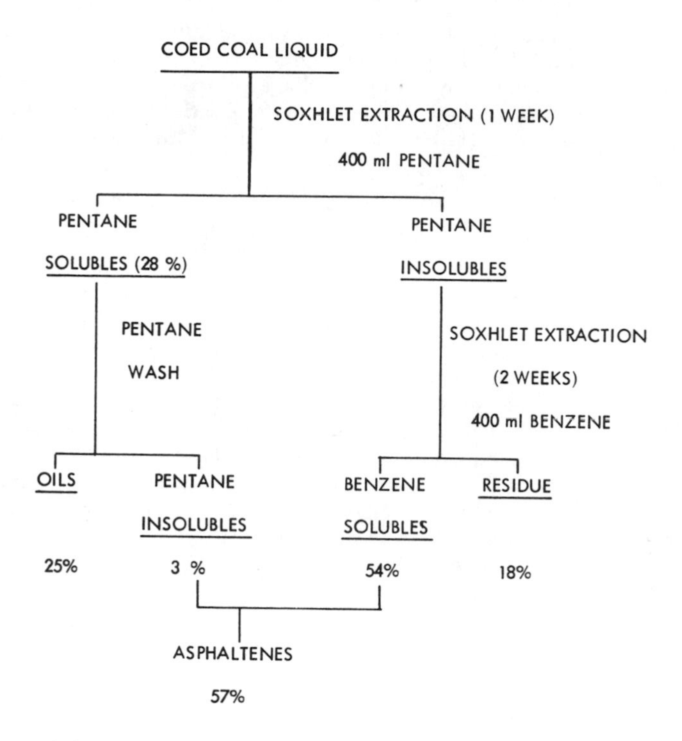

Figure 1. Solvent separation of a COED coal liquid into oils, asphaltenes, and residue

acquired in real time using a Kratos DS-50SM data-acquisition system. High-resolution FI/MS were oscillographically recorded at a dynamic resolution sufficient to resolve two ions differing in mass by about 1 part in 10,000 and 20,000 with approximately 95% and 70% valleys, respectively. With the Kratos instrument, low-voltage EI/MS were obtained at a dynamic resolution of approximately 40,000 in real time using a Kratos DS-50 data acquisition system. Samples were introduced into the ion sources of both mass spectrometers via all-glass inlet systems at temperatures of 300°–320° C.

Data Analysis. Infrared spectra, reproduced in Reference *35*, were qualitatively interpreted using compilations of vibrational frequencies for functional groups in organic substances (*30, 38–43*). Calculation of functional-group concentrations using average extinction coefficients and average molecular weights (*30, 39, 40, 41*) was not attempted.

Precise molecular-ion masses obtained by both 70-eV and low-voltage electron-impact-induced ionization revealed the presence of compounds adhering to the general formula $C_nH_{2n+Z}N_aO_bS_c$ where $Z(H,N_a,O_b,S_c)$ is the specific-Z value (compound type) and $0 \leq a \leq 1$, $0 \leq b \leq 4$, and $0 \leq c \leq 1$. Equation 1 expresses the weight percent (WP) of the ith homolog in the jth specific-Z series, designated simply as $(Z_j)_i$, in terms of both the molecular-ion intensities, $I(Z_j)_i$, and the gram sensitivities, $s(g_j)_i$, for ionization by either high-electric fields or low-voltage electrons. In Equation 1, $Z_j = n - 2(j)$ for $j = 1,2,3 \ldots, l$ for each

$$WP(Z_j)_i = \frac{I(Z_j)_i/s(g_j)_i}{\sum_{j=1}^{l}\sum_{i=1}^{n} I(Z_j)_i/s(g_j)_i} \times 100 \qquad (1)$$

value of a,b, and c, and $n = 4$ or 3 for even or odd values of a, respectively. The available data indicate that the mole sensitivities, $s(m_j)_i$, for ionization of aromatic compounds of present interest by both low-voltage electrons (*44*, *45*) and high-electric fields (*46*), are essentially constant within a specific-Z series. Thus, substitution of the expression $s(g_j)_i = MW(Z_j)_i/s(m_j)$ into Equation 1 leads to Equation 2.

$$WP(Z_j)_i = \frac{I(Z_j)_i \cdot MW(Z_j)_i/s(m_j)}{\sum_{j=1}^{l}\sum_{i=1}^{n} I(Z_j)_i \cdot MW(Z_j)_i/s(m_j)} \times 100 \quad (2)$$

The weight percents of the individual homologs in each specific-Z series (carbon-number distributions) were calculated from LV/EI/MS molecular-ion intensities assuming constant mole sensitivities for each specific-Z series. An invalid factor was inadvertently used in the previous conversion of the LV/EI/MS carbon-number distributions for the asphaltene neutral fraction to carbon-number distributions based on the total liquid. Consequently, the entries in the LV/EI/MS carbon-number distributions for Z(H), Z(O), and Z(S) asphaltene neutral aromatic compounds in References *35* and *47*, the total weight percentages of these specific-Z series in References *35*, *47*, and *48*, and the sums of these latter weight percentages reported in all these references should be multiplied by 0.892.

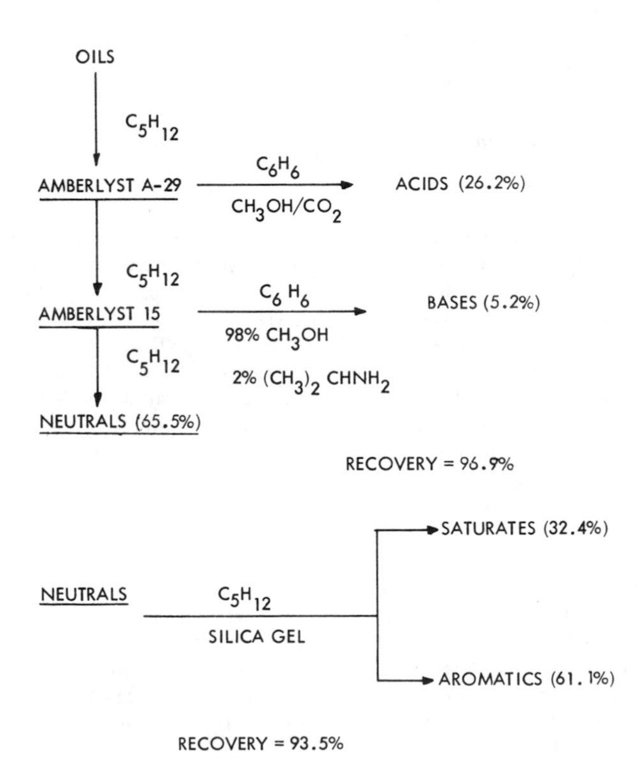

Figure 2. Chromatographic separation of oils

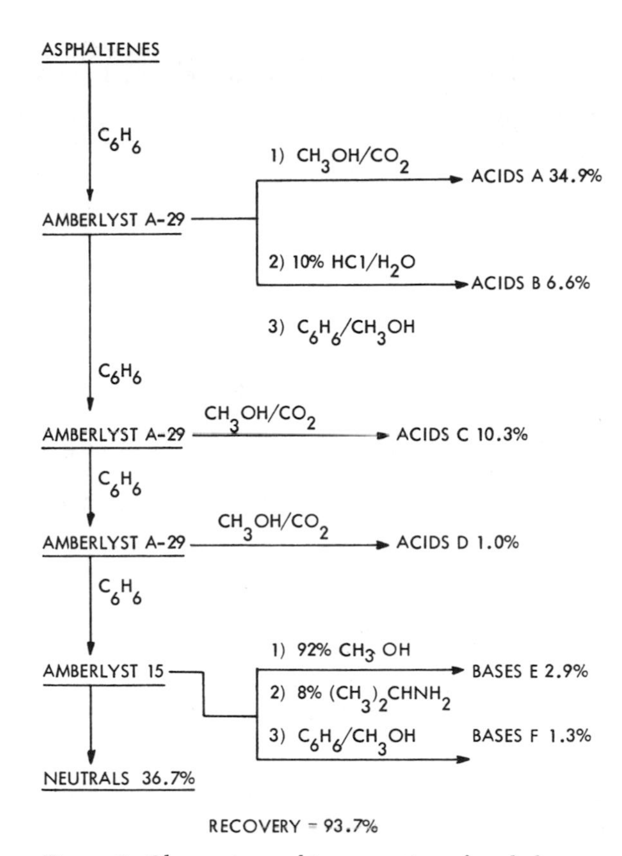

Figure 3. Chromatographic separation of asphaltenes

The FI/MS molecular-ion intensities for both the oil and asphaltene neutral aromatic fractions were converted to carbon-number distributions assuming equal relative mole sensitivities for all specific-Z series and using average relative mole sensitivities for each specific-Z series calculated from either the experimental relative mole sensitivities for individual compounds or from the correlation of these values with specific-Z values (*46*). Comparison of the weight percentages in Columns 2 and 3, and 5 and 6 in Table I is consistent with the suggestion (*46*) that use of the latter approximation in processing FI/MS ion intensities results in acceptable quantitative distributions for fractions comprised of related compound types. Thus, the FI/MS data for all oil and asphaltene acid and base fractions were converted to quantitative distributions assuming unit relative mole sensitivities.

The carbon-number distributions for aromatic fractions obtained by both ionizing techniques are tabulated in References *35* and *47*. As previously noted, the weight percentages by LV/EI/MS should be multiplied by 0.892.

Results and Discussion

General. Table II presents the results (on a total-liquid basis) obtained from separation of the oils and asphaltenes into acids, bases, and neutrals and

Table I. Effect of Sensitivity Approximations on Quantitative Distributions for Asphaltene Neutrals by FI/MS.

	Weight Percents			*Weight Percents*	
$-Z\,(H)$	$s(m_j) = 1$[a]	$s(m_j)$[b]	$-Z\,(O)$	$s(m_j) = 1$[a]	$s(m_j)$[b]
6	3.6	5.5	8	1.8	2.6
8	4.8	6.8	10	2.7	3.8
10	5.6	7.6	12	3.2	3.8
12	10.9	13.1	14	2.5	2.7
14	9.7	10.8	16	5.4	5.8
16	9.8	10.7	18	5.1	5.0
18	12.2	12.6	20	5.1	4.9
20	8.6	8.1	22	3.5	3.5
22	12.0	11.6	24	3.9	3.6
24	8.4	7.8	26	3.6	3.2
26	9.5	8.4	28	3.0	2.4
28	5.2	4.3	30	2.0	1.6
30	6.9	5.6	32	1.4	1.2
32	3.0	2.6			
34	3.5	2.7			
36	1.5	1.2			

[a]Calculations assume unit relative mole sensitivities.
[b]Calculations assume constant relative mole sensitivities within a specific-Z series.

the percentage of each fraction volatile in the batch-inlet system of each mass spectrometer at 300°–320° C. Table II and Figures 2 and 3 reveal that 92.6% and 93.7% of the oils and asphaltenes, respectively, were accounted for in the separation. Although not quantitative, the recovery was considered acceptable for our purposes since at least part of the nonquantitative recoveries were attributable to losses in workup of eluted fractions.

The volatilities for the various asphaltene fractions in the two inlet systems agree within the limits of experimental accuracy. Previously reported (*35, 47, 48*) percentage volatilities for asphaltene-acid-fraction **C** are incorrect; the present values were used in placing both quantitative analyses on a total-liquid basis.

The agreement between the two analyses regarding the percentage saturates in the asphaltene neutrals is satisfactory as follows. The Oklahoma State University (OSU) value of 3.9% was obtained from silica gel chromatography of the pentane-soluble (88.2%) asphaltene neutrals and, hence, represents a lower limit for the percentage saturates in the asphaltene neutrals. The value of 5.2% results from a proprietary Exxon saturate/aromatic hydrocarbon EI/MS technique developed for low-boiling coal liquids.

Oil and Asphaltene Aromatic Neutrals. Tables III and IV summarize the compositional data for the C_nH_{2n+Z} and $C_nH_{2n+Z}O$ compounds present in the oil and asphaltene neutrals. In addition, LV/EI/MS analysis using direct

Table II. Summary of Separation Data and Percent Volatilities for Oil and Asphaltene Fractions

	Oils		*Asphaltenes*			
					Percent Volatile	
Class	*Weight Percent*[a]	*Percent Volatile 21-110B*	*Fraction*	*Weight Percent*[a]	*21-110B*	*MS-50*
Acids	6.6	100	**A**	19.9	50	47
			B	3.8	31	32
			C	5.9	62	66
			D	0.6	44	50
			total	30.2		
Bases	1.3	100	**E**	1.6	77	80
			F	0.7	63	59
			total	2.3		
Neutrals					97	97
saturates	5.3			3.9[b]		
				(5.2)[c]		
aromatics	10.0	100		17.0		
				(15.0)[c]		
total	23.2		total	53.4		
Sample Loss	1.8			3.6		

[a] Based on total liquid.
[b] From silica gel chromatography.
[c] From MS analysis of volatiles.

Table III. Summary Data for C_nH_{2n+Z} Aromatic Hydrocarbon Types

		Oils[a]			*Asphaltenes*[a]			*Asphaltenes*[b]		
			Range In			*Range In*			*Range In*	
−Z(H)	*Parent Formula*	*Weight Percent × 10*[1]	*N*	*Molecular Weight*	*Weight Percent × 10*[1]	*N*	*Molecular Weight*	*Weight Percent × 10*[1]	*N*	*Molecular Weight*
6	C_6H_6	9.7	9–29	120–400	5.5	7–31	92–428	3.7	7–35	92–484
8	C_9H_{10}	12.2	10–29	132–398	6.8	9–37	118–510	6.8	9–41	118–566
10	C_9H_8	10.4	9–30	116–410	7.6	9–40	116–550	6.0	9–40	116–550
12	$C_{10}H_8$	21.1	10–30	128–408	13.1	10–40	128–548	10.9	10–43	128–590
14	$C_{12}H_{10}$	12.9	12–30	154–406	10.8	12–40	154–546	8.7	12–41	154–560
16	$C_{12}H_8$	9.5	12–30	152–404	10.7	12–40	152–544	9.8	12–41	152–558
18	$C_{14}H_{10}$	9.7	14–30	178–402	12.6	14–40	178–542	11.2	14–41	178–556
20	$C_{15}H_{10}$	3.1	15–30	190–400	8.1	15–40	190–540	7.8	15–44	190–596
22	$C_{16}H_{10}$	3.6	16–30	202–398	11.6	16–41	202–552	11.1	16–40	202–538
24	$C_{18}H_{12}$	1.5	18–31	228–410	7.8	18–41	228–550	6.8	18–42	228–564
26	$C_{18}H_{10}$	1.0	18–31	226–408	8.4	18–41	226–548	8.1	19–42	230–562
28	$C_{20}H_{12}$				4.3	20–39	252–518	3.6	20–42	252–560
30	$C_{22}H_{14}$				5.6	22–40	278–530	8.4	22–42	278–558
32	$C_{22}H_{12}$				2.6	22–39	276–514	1.6	23–43	290–570
34	$C_{24}H_{14}$				2.7	22–41	274–540	0.4	24–40	302–526
36	$C_{24}H_{12}$				1.2	24–41	300–538	0.3	24–43	300–566
38	$C_{26}H_{14}$							0.3	26–41	326–536
40	$C_{28}H_{16}$							0.2	28–39	352–508
42	$C_{28}H_{14}$							0.04	29–39	364–506
Total		94.7			119.4			105.7		

[a]Results from FI/MS analysis.
[b]Results from LV/EI/MS analysis.

Table IV. Summary Data for $C_nH_{2n+Z}O$ Neutral Aromatic Compound Types

		Oils[a]			*Asphaltenes*[a]			*Asphaltenes*[b]		
			Range In			*Range In*			*Range In*	
−Z(O)	*Parent Formula*	*Weight Percent × 10¹*	*N*	*Molecular Weight*	*Weight Percent × 10¹*	*N*	*Molecular Weight*	*Weight Percent × 10¹*	*N*	*Molecular Weight*
8	C_8H_8O	1.4	10–17	148–246	2.6	8–19	120–274	3.3	8–38	120–540
10	C_8H_6O	2.8	9–17	132–244	3.8	8–35	118–496	3.4	9–41	132–580
12	$C_{10}H_8O$	2.0	11–20	158–284	3.8	10–32	144–452	2.7	10–42	144–592
14	$C_{12}H_{10}O$	0.4	15–17	212–240	2.7	12–31	170–436	2.4	12–41	170–576
16	$C_{12}H_8O$	2.7	12–20	168–280	5.8	12–39	168–546	3.8	12–41	168–574
18	$C_{14}H_{10}O$	1.5	14–23	194–320	5.0	14–38	194–530	3.3	14–37	194–516
20	$C_{14}H_8O$				4.9	15–38	206–528	2.9	15–39	206–542
22	$C_{16}H_{10}O$				3.5	16–27	218–372	3.1	16–40	218–554
24	$C_{17}H_{10}O$				3.6	18–36	230–496	2.7	17–40	216–552
26	$C_{18}H_{10}O$				3.2	18–39	242–536	2.3	18–41	242–564
28	$C_{20}H_{12}O$				2.4	20–40	268–548	1.9	20–40	268–548
30	$C_{21}H_{12}O$				1.6	21–40	280–546	1.3	21–42	280–574
32	$C_{22}H_{12}O$				1.2	23–39	306–530	0.9	23–39	306–530
34	$C_{24}H_{14}O$							0.5	24–39	318–528
36	$C_{24}H_{12}O$							0.4	24–35	316–470
Total		10.8			44.1			35.0		

[a]Results from FI/MS analysis.
[b]Results from LV/EI/MS analysis.

and indirect methods (*49*) detected approximately 0.7% $C_nH_{2n+Z}S$ compounds ($-30 \leq Z(S) \leq -10$) in the asphaltene neutrals. For the mass range of interest, direct identification of these compound types in the FI/MS–70-eV EI/MS analysis is generally precluded by the resolution capability of the 21-110B (*49*). Since we did not use indirect methods (*49*) to account for aromatic sulfur-containing compounds in the OSU analysis, the weight percents for these Z(S) compounds are included in the values for aromatic hydrocarbons at $-Z(H) = -Z(S) + 4$. However, the error so introduced into the OSU analyses will be negligible because the asphaltene neutral aromatic sulfur-containing compounds account for only approximately 0.7% of the total liquid. The perturbation in the analytical data for the oils must also be minimal as follows. The sulfur content of the total liquid is 0.35% by elemental analysis. From the Exxon analysis the sulfur compounds in the volatilized asphaltene fractions are about 0.8%. Using a value of 300 for the average molecular weight of the COED volatiles as indicated by the MS analyses, the percentage sulfur-containing compounds in the oils must be less than 3.3% because sulfur-containing compounds are undoubtedly present in the nonvolatile asphaltenes and in the 18% benzene insolubles. Thus, the significant conclusions concerning both the agreement between the two MS analyses and the detailed molecular compositions of the oils and asphaltenes are unaltered even though these nonunique difficulties with the analysis of sulfur-containing compounds (*49*) will, all other factors being equal, distort the carbon-number distributions for at least the aromatic hydrocarbons. Consequently, further efforts concerning the analysis of these compounds were not attempted.

The IR spectra of the oil and asphaltene neutrals (*35*) exhibited no significant absorptions in the region 3200–3600 cm^{-1} except for H_2O bands at 3620–3695 cm^{-1} in the matrix-isolation spectrum of the oil neutrals. Weak absorptions near 1700 cm^{-1} are indicative of minor amounts of ketones/aldehydes in both neutral fractions. The absorptions at 2860, 2950, and 3050 cm^{-1} are ascribable to aliphatic and aromatic CH stretching. The band at 1600 cm^{-1} is characteristic of aromatic ring C=C. Thus, the oxygen-containing compounds in both neutral fractions are principally composed of ethers.

For both the oil and asphaltene neutrals, Tables III and IV present in Columns 4, 7, and 10 the range in carbon numbers; in Columns 5, 8, and 11 the range in molecular weights; and in Columns 3, 6, and 9 the weight percents of the total liquid accounted for by each specific-Z series in Column 1. The results in Columns 6–8 and in Columns 9–11 in both tables were calculated from FI/MS and LV/EI/MS data, respectively, assuming constant mole sensitivities in a specific-Z series.

Comparison of the data in Columns 6–8 with the corresponding entries in Columns 9–11 in Table III and in Table IV reveals that the two independent analyses are in excellent qualitative and quantitative agreement regarding the composition of the asphaltene neutrals. The carbon-number distributions for each Z(H) and Z(O) series generally extend to higher molecular weights in the Exxon than in the Oklahoma State analysis. In addition, −38(H) through −42(H), −34(O), and −36(O) compounds were detected in the former but not in the latter analysis. However, the contribution of these compounds to the Exxon data is only 0.16% for the hydrocarbons and 0.33% for the oxygenates (*35*). Furthermore, homologs in the −8(O) and −22(O) series account for 0.20% (61%) of the latter value. These differences undoubtedly reflect the enhanced sensitivity of the MS-50 compared with the 21-110B.

Table III reveals that both the oils and asphaltenes contain appreciable quantities of −6(H) through −26(H) aromatic hydrocarbons. Except for the −6(H) and −8(H) series, the carbon-number distributions for these hydrocarbon types commence at the same molecular weights in both fractions but extend to higher molecular weights in the asphaltenes than in the oils. Both analyses reveal that: the −6(H) and −8(H) series commence at one and two carbon numbers lower in the asphaltenes than in the oils, respectively; and there is an appreciable overlap in the compositions of these two fractions. The latter result is at least qualitatively consistent with the observation that the asphaltene neutrals are about 88% pentane soluble. However, both results are clearly inconsistent with a separation based upon solubilities producing compositionally unique fractions. In this regard, both the oils and asphaltenes contain appreciable quantities of naphthalenes, −12(H), acenaphthenes/biphenyls, −14(H), fluorenes/acenaphthylenes, −16(H), and phenanthrenes/anthracenes, −18(H). The oils and asphaltenes have aromatic hydrocarbons possessing 1–4 and 1–7 aromatic rings, respectively.

The carbon-number distributions for the aromatic hydrocarbons in the oils and asphaltenes (*35*) are characterized by the occurrence of the maximum weight percent in the early part of each Z(H) series. In this regard, it should be noted that the COED process involves multi-stage pyrolysis. Thus, for aromatic hydrocarbon types, the dependence of the weight percents on carbon number for the COED liquid parallels the dependence observed for liquids obtained from other coal-pyrolysis processes (*50*, *51*, *52*). It is interesting to note that the weight percents for $C_{25}H_{24}$ and $C_{26}H_{30}$ in both the oils and asphaltenes and $C_{24}H_{18}$ in the asphaltenes are significantly larger than would be expected (*35*); the result was observed in both asphaltene analyses. Although the origin of this phenomenon is unclear, it presumedly reflects some combination of coal structure and preferential formation/diminished reactivity in the liquefaction process.

The IR data, as discussed above, and the molecular formulas for the first homologs in at least the less-negative Z(O) series in Table IV, indicate that furans dominate the oxygen-containing compounds in both the oil and asphaltene neutral fractions. Furthermore, comparison of the Z(H) and Z(O) values in Tables III and IV, respectively, suggests that the furans are phenomenologically derived from the aromatic hydrocarbons by replacement of ring CH_2 by O.

The oil and asphaltene neutrals contain −8(O) through −18(O) compounds. Table IV reveals that the carbon-number distributions for the −8(O), −10(O), −12(O), and −14(O) series commence at lower molecular weights in the asphaltenes than in the oils. However, the Z(O) series extend to higher molecular weights in the asphaltenes than in the oils. This result parallels that obtained for the aromatic hydrocarbons. The oil and asphaltene neutral fractions contain aromatic oxygen-containing compounds with 2–3 and 2–7 aromatic rings, respectively. Both fractions have appreciable amounts of dihydrobenzofurans, −8(O); benzofurans, −10(O); naphthenobenzofurans, −12(O); dibenzofurans, −16(O); and acenaphthenofurans, −18(O).

The carbon-number distributions (35) for the $C_nH_{2n+Z}O$ compounds in the oil neutrals are relatively short and exhibit maxima near the center of each Z(O) series. In contrast, the carbon-number distributions for the asphaltene-neutral Z(O) compounds are characterized by the occurrence of maximums in the beginning of each distribution. The origin of the differences in these Z(O) distributions is not clear. Furthermore, the former result contrasts with the result obtained for the oil aromatic hydrocarbons and the latter parallels the distribution of weight percents in each Z(H) series for the asphaltene neutrals. It is tempting to ascribe the difference in the distribution of weight percents of the various Z(O) compound types in the oil and asphaltene neutrals to the chemistry associated with the COED process. However, the fact that considerable overlap exists in the neutral-aromatic Z(O) compounds between the oils and asphaltenes precludes this conclusion because the results could reflect phenomena associated with solvent extraction.

Oil and Asphaltene Acids. Tables V and VI contain compositional data for the $C_nH_{2n+Z}O$ and $C_nH_{2n+Z}O_2$ compounds present in the oil and four asphaltene-acid fractions. Analytical data for compounds in these fractions containing either three oxygens, four oxygens, a nitrogen, or a nitrogen and oxygen are summarized in Table VII. In regard to the composition of asphaltene-acid-fraction **C**, both MS analyses determined the presence of −6 through −34 Z(O) and $Z(O_2)$ compounds. Minor amounts of −36(O) and $-36(O_2)$ compounds were observed by LV/EI/MS but not by FI/MS. Tables V and VI also show that the two analyses of acid-fraction **C** differ slightly in terms of the total number of carbons in a given specific-Z series. However, the Z(O) and $Z(O_2)$ carbon-number distributions (35) reveal that this difference reflects the presence of compounds accounting for only approximately 0.3% of the total liquid.

As seen in Table VII, both analyses detected $Z(O_3)$, $Z(O_4)$, Z(N), and Z(N,O) compounds in acid-fraction **C**. However, the summary data for these compounds by LV/EI/MS are presented because resolution of $C_{n-1}{}^{13}CH_{2n+Z}O$ molecular ions from $C_nH_{2n+Z}NO$ molecular ions and of $C_{n-1}{}^{13}CH_{2n+Z}O_3$ molecular ions from $C_nH_{2n+Z}N$ molecular ions was generally precluded in the FI/MS analysis. Thus, analysis of these compound types in the latter experiments reflects the magnitude of the monoisotopic intensities (53) at even and odd *m/e* values. In addition to $Z(O_3)$ and $Z(O_4)$ compounds, the

FI/MS analysis detected the presence of small amounts of Z(N) compounds in all acid fractions and Z(N,O) compounds in acid-fraction **D**.

For asphaltene-acid-fraction **C**, the volatile compound types in Table VII account for about 0.4% and 0.2% of the total liquid by LV/EI/MS and FI/MS, respectively. This fraction was indicated to be slightly more volatile in the 21-110B-inlet system than in the MS-50-inlet system (*see* Table II). Thus, the two analyses are in good agreement concerning the weight percentage of the volatile compounds in acid-fraction **C** containing 1–4 oxygens, a nitrogen, or a nitrogen and oxygen. Finally, both analyses detected trace amounts of hydrocarbon contaminants in all fractions.

The IR spectra of the oil- and asphaltene-acid fractions (35) exhibit an intense absorption band at 3600 cm^{-1} characteristic of phenolic OH stretching. In addition, the matrix-isolation IR spectra of the oil acids and acid-fractions **A** and **C** but not the solution IR spectrum of acid-fraction **D** show broad bands centered near 3300 cm^{-1}. These results, combined with the absence of a strong absorption in the 1700–1750 cm^{-1} region, are explained by intermolecular hydrogen bonded, ring-hydroxylated aromatics in the N_2 matrix. Thus, the IR data for the oil acids and asphaltene-acids **A**, **C**, and **D** confirm the presence of aromatic-ring hydroxylated compounds and exclude carboxylic acids as major contributors to acids possessing at least two oxygens.

In contrast, the IR spectrum of asphaltene-acid-fraction **B** exhibits intense bands near 1700 cm^{-1} and 3200 cm^{-1} (broad band, approximately 2500–3500 cm^{-1}) characteristic of aromatic carboxylic acids. In this regard, the FI/MS of this fraction confirms the existence of molecular ions at m/e 122 and 136 in the $-8(O_2)$ and at m/e 172 in the $-14(O_2)$ series. Indanediols and tetralindiols are parent members in the $-8(O_2)$ series. Since these compound series commence at molecular weights 150 and 164, respectively, benzoic-acid molecular ions and variously methylated homologs rationally explain the ions at m/z 122 and 136. Typical members of the $-14(O_2)$ series are acenaphthalenediols and tetrahydrophenanthrenediols commencing at mass 186 and 214, respectively. Thus, naphthalenoic acid appears to best explain the m/z 172 ion. Any attempt to identify other plausible aromatic carboxylic acids in acid-fraction **B** is precluded by the number of possible isomers at increasing carbon numbers across the various $Z(O_2)$ series.

FI/MS confirms the presence of molecular ions at m/z 122 and 172 in asphaltene-acid-fraction **A** and at m/z 136 in asphaltene-acid-fractions **A** and **D**. As previously noted, the IR spectra preclude the existence of significant amounts of carboxylic acids in these fractions. Assuming the validity of these assignments, it should be noted that benzoic acid in **A** and **B**, singly methylated benzoic acids in **A**, **B**, and **D**, and naphthalenoic acid in **A** and **B** account for only 0.01%, 0.01%, and 0.06% of the total coal liquid, respectively.

To our knowledge, this result represents one of the first observations of carboxylic acids in coal-derived liquids. Either the chemistry associated with the COED process, the nature of the feed coal, or the breakdown of the **A-29**

Table V. Summary Data for $C_nH_{2n+Z}O$ Aromatic Acids

		Oils[a]			*Asphaltene-Fraction* **A**[a]			*Asphaltene-Fraction* **B**[a]		
			Range In			*Range In*			*Range In*	
$-Z(O)$	*Parent Formula*	*Weight Percent* $\times 10^1$	*N*	*Molecular Weight*	*Weight Percent* $\times 10^1$	*N*	*Molecular Weight*	*Weight Percent* $\times 10^1$	*N*	*Molecular Weight*
6	C_6H_6O	13.5	6–23	94–332	5.8	6–11	94–164	0.1	6–10	94–150
8	$C_9H_{10}O$	10.1	9–24	134–344	0.7	9–13	134–190	0.6	9–14	134–204
10	C_9H_8O	4.5	9–24	132–342	1.7	9–16	132–230	0.2	9–14	132–202
12	$C_{10}H_8O$	4.5	10–24	144–340	5.8	10–18	144–256	0.3	10–15	144–214
14	$C_{12}H_{10}O$	4.2	12–24	170–338	1.6	12–15	170–212	0.4	12–18	170–254
16	$C_{12}H_8O$	2.9	12–24	168–336	2.8	12–18	182–252	0.4	12–19	182–266
18	$C_{14}H_{10}O$	4.1	14–24	194–334	3.4	14–21	194–292	0.2	14–22	194–306
20	$C_{15}H_{10}O$	1.4	15–24	206–332	2.6	15–22	206–304	0.2	15–21	206–290
22	$C_{16}H_{10}O$	1.3	16–25	218–344	2.1	16–22	218–302	0.8	16–24	218–330
24	$C_{17}H_{10}O$				1.9	17–24	230–328	0.3	17–24	230–328
26	$C_{18}H_{10}O$				2.5	18–24	242–326	0.2	19–25	256–340
28	$C_{20}H_{12}O$				2.2	20–26	268–352	0.2	20–27	268–366
30	$C_{21}H_{12}O$				1.5	21–26	280–350	0.1	22–29	294–392
32	$C_{22}H_{12}O$				0.9	22–27	292–362	0.1	24–27	320–362
Total		46.5			35.5			4.1		

Table V. Continued.

		Asphaltene-Fraction C[a]			*Asphaltene-Fraction* C[b]			*Asphaltene-Fraction* D[a]		
			Range In			*Range In*			*Range In*	
$-Z(O)$	*Parent Formula*	*Weight Percent* $\times 10^1$	*N*	*Molecular Weight*	*Weight Percent* $\times 10^1$	*N*	*Molecular Weight*	*Weight Percent* $\times 10^1$	*N*	*Molecular Weight*
6	C_6H_6O	4.4	6–14	94–206	5.9	6–14	94–206	0.06	6–13	94–192
8	$C_9H_{10}O$	3.4	9–15	134–218	2.2	8–16	120–232	0.04	9–16	134–232
10	C_9H_8O	1.4	9–16	132–230	0.8	9–19	132–272	0.08	9–16	132–230
12	$C_{10}H_8O$	1.2	10–17	144–242	1.1	10–21	144–298	0.08	10–18	144–256
14	$C_{12}H_{10}O$	1.5	12–18	170–254	1.0	12–21	170–296	0.05	12–18	170–254
16	$C_{12}H_8O$	3.3	13–20	182–280	1.9	13–26	182–364	0.07	13–20	182–280
18	$C_{14}H_{10}O$	1.9	14–21	194–292	1.3	14–29	194–404	0.06	14–23	194–320
20	$C_{15}H_{10}O$	1.4	15–22	206–304	1.2	14–29	192–402	0.06	15–23	206–318
22	$C_{16}H_{10}O$	2.1	16–25	218–344	2.0	16–29	218–400	0.08	16–25	218–344
24	$C_{17}H_{10}O$	1.5	17–25	230–342	1.4	17–31	230–426	0.07	18–26	244–356
26	$C_{18}H_{10}O$	1.3	18–26	242–354	1.1	18–32	242–438	0.05	19–26	256–354
28	$C_{20}H_{12}O$	0.9	20–26	268–352	0.7	20–31	268–422	0.04	20–26	268–352
30	$C_{21}H_{12}O$	0.6	24–26	322–350	0.5	21–31	280–420	0.05	22–28	294–378
32	$C_{22}H_{12}O$	0.2	24–26	320–348	0.2	22–33	292–446	0.02	22–26	292–348
34	$C_{24}H_{14}O$	0.2	25–27	332–360	0.1	24–33	318–444	0.01	25–27	332–360
36	$C_{24}H_{12}O$				0.1	24–30	316–400			
Total		25.3			21.5			0.8		

[a]Results from FI/MS analysis.
[b]Results from LV/EI/MS analysis.

Table VI. Summary Data for $C_nH_{2n+Z}O_2$ Aromatic Acids

		Oils[a]			Asphaltene-Fraction A[a]			Asphaltene-Fraction B[a]		
			Range In			Range In			Range In	
−Z(O)	Parent Formula	Weight Percent × 10^1	N	Molecular Weight	Weight Percent × 10^1	N	Molecular Weight	Weight Percent × 10^1	N	Molecular Weight
6	$C_6H_6O_2$	1.7	6–11	110–180	7.6	6–13	110–208	1.2	6–13	110–208
8	$C_9H_{10}O_2$	1.5	9–22	150–332	1.1	7–13	122–206	0.2	7–11	122–178
10	$C_9H_8O_2$	2.9	9–23	148–344	2.9	9–16	148–246	0.5	9–16	148–246
12	$C_{10}H_8O_2$	1.8	10–23	160–342	0.8	10–14	160–216	0.4	10–16	160–244
14	$C_{12}H_{10}O_2$	2.4	12–23	186–340	4.5	11–19	172–284	0.3	11–16	172–242
16	$C_{12}H_8O_2$	3.2	12–23	184–338	9.8	12–20	184–296	0.4	12–20	184–296
18	$C_{14}H_{10}O_2$	3.5	14–23	210–336	5.0	14–23	210–336	0.3	14–21	210–308
20	$C_{15}H_{10}O_2$				2.1	15–24	222–348	0.3	15–23	222–334
22	$C_{16}H_{10}O_2$				2.9	16–25	234–360	0.3	16–24	234–346
24	$C_{17}H_{10}O_2$				2.5	17–26	246–372	0.3	17–23	246–330
26	$C_{18}H_{10}O_2$				2.2	18–26	258–370	0.2	18–25	258–356
28	$C_{20}H_{12}O_2$				1.4	20–27	284–382	0.2	20–27	284–382
30	$C_{21}H_{12}O_2$				0.9	21–27	296–380	0.1	21–27	296–380
32	$C_{22}H_{12}O_2$				0.9	22–28	308–392	0.1	22–26	308–364
34	$C_{24}H_{14}O_2$				0.4	24–29	334–404	0.04	24–26	334–362
36	$C_{24}H_{12}O_2$				0.2	25–28	346–388	0.1	24–27	332–374
Total		17.0			45.2			4.9		

Table VI. Continued

		Asphaltene-Fraction C[a]			Asphaltene-Fraction C[b]			Asphaltene-Fraction D[a]		
			Range In			Range In			Range In	
−Z(O)	Parent Formula	Weight Percent × 10^1	N	Molecular Weight	Weight Percent × 10^1	N	Molecular Weight	Weight Percent × 10^1	N	Molecular Weight
6	$C_6H_6O_2$	0.2	6–10	110–166	0.2	6–10	110–166	0.04	6–11	110–180
8	$C_9H_{10}O_2$	0.2	9–13	150–206	0.1	9–13	150–206	0.04	8–12	136–192
10	$C_9H_8O_2$	0.7	9–14	148–218	0.8	9–18	148–274	0.04	9–16	148–246
12	$C_{10}H_8O_2$	0.4	10–16	160–244	0.3	10–18	160–272	0.05	10–18	160–272
14	$C_{12}H_{10}O_2$	0.5	12–19	186–284	0.2	12–22	186–326	0.04	12–18	186–270
16	$C_{12}H_8O_2$	1.2	12–19	184–282	1.1	12–26	184–380	0.05	13–19	198–282
18	$C_{14}H_{10}O_2$	1.2	14–21	210–308	0.9	14–28	210–406	0.09	14–21	210–308
20	$C_{15}H_{10}O_2$	1.1	16–23	236–334	0.9	15–28	222–404	0.05	17–22	250–320
22	$C_{16}H_{10}O_2$	1.3	16–23	234–332	1.1	16–28	234–402	0.05	16–24	234–346
24	$C_{17}H_{10}O_2$	1.6	17–26	246–372	1.2	18–32	260–456	0.09	17–26	246–372
26	$C_{18}H_{10}O_2$	0.9	19–26	272–370	1.0	19–31	272–440	0.06	20–26	286–370
28	$C_{20}H_{12}O_2$	0.5	20–26	284–368	0.6	20–32	284–452	0.05	20–25	284–354
30	$C_{21}H_{12}O_2$	0.5	22–26	310–366	0.6	22–32	310–450	0.03	22–25	310–352
32	$C_{22}H_{12}O_2$	0.4	25–27	350–378	0.4	22–32	308–448	0.04	23–27	322–378
34	$C_{24}H_{14}O_2$	0.04	25	348	0.2	24–33	334–460	0.01	24	334
36	$C_{24}H_{12}O_2$				0.02	25–33	346–458	0.01	25–26	346–360
Total		10.7			9.6			0.7		

[a] Results from FI/MS analysis.
[b] Results from LV/EI/MS analysis.

Table VII. Summary Data for Other Aromatic Acids

Fraction	*General Formula*	*Specific-Z Series*	*Weight Percent* × 10^{1}
Oils[a]	$C_nH_{2n+Z}N$	−15 (N)	1.1
Asphaltene-Fraction **A**[a]	$C_nH_{2n+Z}O_3$	−16 through −32 (O_3)	10.0
	$C_nH_{2n+Z}O_4$	−16, −18, −26, and −28 (O_4)	3.2
Asphaltene-Fraction **B**[a]	$C_nH_{2n+Z}O_3$	−14 through −30 (O_3)	1.1
	$C_nH_{2n+Z}O_4$	−18 through −28 (O_4)	0.4
Asphaltene-Fraction **C**[b]	$C_nH_{2n+Z}N$	−5 through −21 (N)	0.6
	$C_nH_{2n+Z}O_3$	−6 through −36 (O_3)	1.5
	$C_nH_{2n+Z}O_4$	−6 through −34 (O_4)	0.7
	$C_nH_{2n+Z}NO$	−5 through −33 (N,O)	1.2
Asphaltene-Fraction **D**[a]	$C_nH_{2n+Z}O_3$	−16 through −32 (O_3)	0.2
	$C_nH_{2n+Z}O_4$	−18 through −28 (O_4)	0.05
	$C_nH_{2n+Z}NO$	−5, −9, −11, −15 through −19 (N,O)	0.4

[a]Results from FI/MS analysis.
[b]Results from LV/EI/MS analysis.

resin during the separation procedure could explain the origin of these acids. Although the present data are inadequate to eliminate the last phenomenon as an explanation, the failure to observe carboxylic acids in asphaltene-acid-fraction **C**, in the oil acids, and in the acid fractions from separation of other coal-derived liquids conducted at OSU (*51*, *52*) renders it somewhat tenuous.

The chemistry associated with ion exchange chromatography (*37*, *38*) and the carbon numbers for at least the first homologs in the less-negative Z(N) series suggest that pyrroles account for the aromatic nitrogen-containing compounds in the various acid fractions. Absorption bands in the 3470-cm^{-1} region are characteristic of pyrrolic NH. The nondescript nature of the 3400-cm^{-1} region in the KBr spectrum of acid-fraction **B**, in the matrix-isolation spectrum of acid-fraction **C**, and in the solution spectrum of acid-fraction **D** (*35*) preclude any conclusions concerning either the presence or the absence of pyrroles in these fractions. In contrast, the matrix-isolation spectra of the oil acids and asphaltene-acid-fraction **A** (*35*) reveal resolved absorption bands in the 3470-cm^{-1} region superimposed on the 2800–3500 cm^{-1} absorption (*see above*). It appears unlikely that these absorptions result from intermolecular hydrogen bonding between water molecules trapped in the N_2 matrix because absorptions for this phenomenon normally occur at 3620–3695 cm^{-1}. Consequently, the IR spectra for these two acid fractions support the presence of pyrroles. The IR data preclude any assessment of the contribution of pyrroles containing either a phenolic group or a furan ring to the NO-containing acids.

As seen in Table II, the four asphaltene-acid fractions exhibit different volatilities. However, the MS data in Tables V–VII and the IR data reveal no significant differences in the group types present in the volatiles from these four fractions. Consequently, the production of four acid fractions represents the use of an inadequate quantity of **A**-29 resin in the asphaltene separation rather than a separation based upon compound type. Therefore, the individual asphaltene-acid fractions could have been combined prior to analysis.

The previous results show that the oxygen-containing acids in both the oils and asphaltenes are principally composed of phenolic types phenomenologically derived from the aromatic hydrocarbons by substitution of 1–4 hydroxyl groups for aromatic hydrogens. The data in Tables V and VI reveal that the oils and asphaltenes contain appreciable numbers of phenolic compound types in the -6 Z(O) through -22 Z(O) and -6 $Z(O_2)$ through -18 $Z(O_2)$ series. This result and the absence in the oils of aromatic acids containing three or four OH groups (*see* Table VII) suggests that the maximum ring number for contamination during solvent separation decreases with increasing hydroxylation of the aromatic ring. In addition, the Z(O) and $Z(O_2)$ acidic compound types common to both the oils and asphaltenes commence at the same molecular weights but extend to higher carbon numbers in the oils than in the asphaltene fractions. This result is especially evident in the -6 Z(O) through -14 Z(O) series and in the -8 $Z(O_2)$ through

-14 Z(O_2) series. It is not obvious how this result, which is opposite to that observed for the aromatic neutral fractions, is reconciled with a separation of acids based upon their physical properties, such as solubilities, in pentane and benzene. Similar to the result attained for the aromatic hydrocarbons, the carbon-number distributions for the acid fractions are characterized by the occurrence of maxima early in the distributions of weight percents in the various specific–Z series. However, for corresponding compound types (i.e., the same basic ring structure) the range in carbon number is significantly shorter for the acids than for the aromatic hydrocarbons.

Oil and Asphaltene Bases. The compositional data for the volatile nitrogen- and nitrogen-plus-oxygen-containing compounds in the oil- and asphaltene-base fractions on a total liquid basis are summarized in Tables VIII and IX, respectively. The compound types in these tables represent the major contributors to the composition of the volatiles from these base fractions. Table X lists minor compound types present in these volatiles. Detection of many of these minor compound types in the Exxon but not in the OSU analysis is attributable to the enhanced sensitivity and resolution of the MS-50 compared with the 21-110B.

Tables VIII and IX demonstrate reasonable agreement between the two analyses in regard to the range in molecular weight for both Z(N) and Z(N,O) volatile bases. The Z(N) and Z(N,O) carbon-number distributions (*35*) reveal that the differences in Tables VIII and IX reflect low-intensity molecular ions at higher *m/z* values in the LV/EI/MS than in the FI/MS. Thus, the good agreement in the weight percents for the various Z(N) and Z(N,O) series between the two analyses is not an artifact.

The oil and asphaltene base fractions are seen to contain aromatic compound types in the -5 Z(N) through -25 Z(N) series and -7 Z(N,O) through -19 Z(N,O) series, respectively. This result suggests that introduction of an oxygen to the aromatic nitrogen-containing bases reduces the maximum ring number for contamination during the solvent separation. The observation that the oils contain compounds in the -11 Z(N,O_2) series as opposed to the asphaltenes, which contain compounds in the -5 Z(N,O_2) through -37 Z(N,O_2) series, supports this conclusion. The carbon-number distributions (*35*) for the Z(N) series common to both the oils and asphaltenes commence at about the same molecular weights in both fractions but extend to slightly higher molecular weights in the asphaltenes than in the oils. In contrast to this result, the carbon-number distributions (*35*) for the Z(N,O) series common to both the oils and asphaltenes commence at about the same molecular weights in both fractions but extend to slightly higher molecular weights in the oils than in the asphaltenes.

The 2500–3600-cm^{-1} region in the IR spectra of the base fractions (*35*) shows no evidence for the presence of OH groups. A weak absorption near 3350 cm^{-1} in the IR spectrum of asphaltene-fraction E could be interpreted as providing tenuous evidence for the presence of basic NH. Weak absorptions

Table VIII. Summary Data for $C_nH_{2n+Z}N$ Aromatic Bases

		Oils[a]			*Asphaltene-Fraction* E[a]			*Asphaltene-Fraction* E[b]			*Asphaltene-Fraction* F[a]		
			Range In			*Range In*			*Range In*			*Range In*	
−*Z(N)*	*Parent Formula*	*Weight Percent × 10*[1]	*N*	*Molecular Weight*	*Weight Percent × 10*[1]	*N*	*Molecular Weight*	*Weight Percent × 10*[1]	*N*	*Molecular Weight*	*Weight Percent × 10*[1]	*N*	*Molecular Weight*
5	C_5H_5N	0.3	8–19	121–275	0.2	8–19	121–275	0.4	7–21	107–303	0.1	8–18	121–261
7	C_8H_9N	0.5	8–18	119–259	0.3	8–20	119–287	0.3	8–22	119–315	0.07	8–20	119–287
9	C_8H_7N	0.6	8–20	117–285	0.3	8–19	117–271	0.3	8–26	117–369	0.08	8–20	117–285
11	C_9H_7N	2.3	9–25	129–353	1.3	9–23	129–325	1.1	9–26	129–367	0.2	9–23	129–325
13	$C_{11}H_9N$	1.1	11–24	155–337	0.7	11–28	155–393	0.6	11–28	155–393	0.2	11–25	155–351
15	$C_{11}H_7N$	0.8	12–25	167–349	0.5	12–23	167–321	0.6	12–27	167–377	0.2	12–23	167–321
17	$C_{13}H_9N$	0.8	13–21	179–291	0.8	13–28	179–389	0.9	13–27	179–375	0.3	13–25	179–347
19	$C_{14}H_9N$	0.6	14–26	191–359	0.4	14–28	191–387	0.3	14–28	191–387	0.2	17–27	233–373
21	$C_{15}H_9N$	0.7	16–26	217–357	0.8	15–28	203–385	0.6	15–28	203–385	0.3	15–29	203–399
23	$C_{16}H_9N$	0.5	16–25	215–341	0.5	17–29	229–397	0.4	17–30	229–411	0.3	16–29	215–397
25	$C_{17}H_9N$	0.3	18–26	241–353	0.2	18–29	241–395	0.3	18–30	241–409	0.1	18–29	241–395
27	$C_{19}H_{11}N$				0.3	19–29	253–393	0.3	19–29	253–393	0.2	19–29	253–393
29	$C_{21}H_{13}N$				0.3	21–29	279–391	0.2	21–29	279–391	0.1	21–29	297–391
31	$C_{21}H_{11}N$				0.2	21–29	277–389	0.2	22–28	291–375	0.1	21–29	277–389
33	$C_{23}H_{13}N$				0.2	23–30	303–401	0.1	23–29	303–387	0.05	23–30	303–401
35	$C_{23}H_{11}N$				0.1	23–30	301–399	0.04	23–29	301–385	0.06	23–30	301–399
37	$C_{25}H_{13}N$				0.03	28–32	269–425	0.02	25–30	327–397	0.05	25–30	327–397
39	$C_{27}H_{15}N$				0.03	27–30	353–395	0.01	27–30	353–395	0.01	27–30	353–395
41	$C_{27}H_{13}N$				0.01	27–29	351–379						
Total		8.5			7.2			6.7			2.6		

[a]Results from FI/MS analysis.
[b]Results from LV/EI/MS analysis.

Table IX. Summary Data for $C_nH_{2n+Z}NO$ Aromatic Bases

		Oils[a]			Asphaltene-Fraction E[a]			Asphaltene-Fraction E[b]			Asphaltene-Fraction F[a]		
			Range In			Range In			Range In			Range In	
−Z(N,O)	Parent Formula	Weight Percent $\times 10^1$	N	Molecular Weight	Weight Percent $\times 10^1$	N	Molecular Weight	Weight Percent $\times 10^1$	N	Molecular Weight	Weight Percent $\times 10^1$	N	Molecular Weight
7	C_7H_7NO	0.5	7–22	121–331	0.1	7–15	121–233	0.1	8–17	135–261	0.04	7–15	121–233
9	C_7H_5NO	0.7	9–23	147–343	0.2	8–16	133–245	0.2	8–20	133–301	0.07	8–15	133–231
11	$C_{10}H_9NO$	0.5	10–23	159–341	1.8	10–19	159–285	1.2	10–22	159–327	0.1	10–18	159–271
13	$C_{10}H_9NO$	0.3	12–24	185–353	0.2	11–18	171–269	0.2	11–25	171–367	0.1	11–18	171–269
15	$C_{11}H_7NO$	0.7	11–23	169–337	0.4	12–20	183–295	0.2	12–27	183–393	0.2	11–21	169–309
17	$C_{13}H_9NO$	0.7	14–24	209–349	0.6	13–22	195–321	0.4	14–27	209–391	0.2	13–22	195–321
19	$C_{13}H_7NO$	1.0	14–25	207–361	0.3	13–21	193–305	0.2	15–27	221–389	0.1	13–23	193–333
21	$C_{15}H_9NO$				0.3	15–22	219–317	0.3	15–27	219–387	0.2	17–24	247–345
23	$C_{16}H_9NO$				0.2	17–26	245–371	0.2	17–27	245–385	0.1	17–24	245–343
25	$C_{18}H_{11}NO$				0.2	18–26	257–369	0.3	18–28	257–397	0.06	18–26	257–369
27	$C_{19}H_{11}NO$				0.2	20–26	283–367	0.3	20–28	283–395	0.1	18–27	255–395
29	$C_{20}H_{11}NO$				0.03	21–25	295–351	0.2	20–28	281–393	0.04	20–28	281–393
31	$C_{21}H_{11}NO$				0.01	23–26	321–363	0.2	20–28	279–391	0.03	24–28	335–391
33	$C_{23}H_{13}NO$				0.02	24–25	333–347	0.1	23–28	319–389	0.02	24–28	333–389
35	$C_{23}H_{11}NO$				0.03	24–25	331–345	0.04	23–28	317–387	0.03	23–29	317–401
37	$C_{24}H_{11}NO$							0.02	24–28	329–385			
39	$C_{26}H_{13}NO$							0.01	26–29	355–397			
41	$C_{27}H_{13}NO$							0.01	27	367			
Total		4.4			4.7			4.2			1.4		

[a] Results from FI/MS analysis.
[b] Results from LV/EI/MS analysis.

Table X. Summary Data for Other Aromatic Bases

Fraction	*General Formula*	*Specific-Z Series*	*Weight Percent × 10¹*
Oils	$C_nH_{2n+Z}NO_2$	-11 (N, O_2)	0.05
Asphaltenes	$C_nH_{2n+Z}NO_2$	-5 through -37(N,O_2)	0.76
	$C_nH_{2n+Z}NS$	-9 through -29 (N, S)	0.13
	$C_nH_{2n+Z}NO_3$, $C_nH_{2n+Z}NS_2$, $C_nH_{2n+Z}NSO$, $C_nH_{2n+Z}NSO_2$, $C_nH_{2n+Z}NS_2O$		trace

near 1700 cm^{-1} in the spectra of all three base fractions indicate the presence of minor amounts of carbonyl groups. Weak absorptions in the 3470-cm^{-1} region of the oil bases could indicate the presence of trace amounts of pyrroles.

Based upon the IR spectra, the method of separation, and the carbon numbers for the first homologs in at least the less negative-Z(N) and -Z(N,O) series, the volatile bases are principally comprised of azaaromatic and azaoxaaromatic compounds. The former and the latter compounds are phenomenologically viewed as arising from replacement of an aromatic-ring CH in the hydrocarbons and furans, respectively, by nitrogen. The basic Z(N) compound types present in the oil and asphaltene volatiles possess 1–5 and 1–7 aromatic rings, respectively. The Z(N,O) compounds present in the volatile oil and asphaltene bases contain 2–4 and 2–5 aromatic rings, respectively.

Summary and Conclusions

The two independent MS analyses are in remarkably good agreement considering that the homologs comprising the aromatic compound types present in the volatile asphaltene acids, bases, and neutrals were identified and quantified using different ionization techniques and mass spectrometers differing significantly in resolution and in the efficiency of ion extraction and transmission. In fact, for the volatile aromatic asphaltene fractions the differences between the weight percents and range in carbon numbers and molecular weights for specific-Z series in Tables III–X are reasonable, recognizing the enhanced capabilities of the MS-50 compared with the 21-110B and the fact that the analyses were conducted independently by two groups. Within these limits, the qualitative and quantitative results are independent of the ionization techniques, the basic mass spectrometer technology, and the groups conducting the analyses. Thus the two MS analyses combined with the IR data lead independently and synergistically to the following conclusions.

The IR and the MS analyses provide detailed support for the contention of Ruberto, Jewell, and Cronauer (*54*) that ion exchange chromatography is suitable for the separation of coal-derived liquids according to compound classes. This result is significant because the applicability of the technique to the separation of coal liquids has been questioned (*55, 56, 57*).

Oils and asphaltenes, which constitute a partial characterization of coal liquids according to solvent extraction, are generally considered to be key intermediates in coal liquefaction. In this regard, the present results do reveal that the asphaltenes contain higher molecular weight homologs in many specific-Z series and different compound types than do the oils. However, our study unequivocally demonstrates that compound types are observed in both the oils and asphaltenes that are equivalent in molecular formula and, hence, presumably in molecular structure. Furthermore, the overlap in the compositions of the two fractions is quantitatively appreciable. Thus, isolation of oils and asphaltenes must involve, in addition to solubility, other physical/

chemical phenomena such as coprecipitation, molecular inclusions, and intermolecular electronic interactions. In fact, Long (33) has suggested that these physical–chemical effects may be quantifiable by developing a polarity scale based upon solubility parameters.

A separation according to solubility would produce oil and asphaltene fractions essentially composed of either unique compound types (specific-Z series) or unique homologs (carbon-number distributions) for corresponding specific-Z series. The implication of this consideration for the characterization of coal liquids according to oils and asphaltenes is illustrated in terms of the analytical data for aromatic hydrocarbons, aromatic acids containing one oxygen, and aromatic nitrogen-containing bases. As a first approximation to a solubility separation, the weight percentages of the homologs in the various specific-Z series for these three compound classes in the asphaltenes that are common to those in the oils were added to the corresponding weight percentages for the latter. The results so obtained from the FI/MS data for the aromatic hydrocarbons, singly-oxygenated aromatic acids, and nitrogen-containing aromatic bases are summarized in Tables XI, XII, and XIII, respectively.

Table XI. Summary Data for Aromatic Hydrocarbons in Compositionally Corrected Oils and Asphaltenes[a]

	Oils			*Asphaltenes*		
		Range In			*Range In*	
$-Z(H)$	*Weight Percent* $\times 10^1$	*N*	*Molecular Weight*	*Weight Percent* $\times 10^1$	*N*	*Molecular Weight*
6	10.7	9–29	120–400	0.04	30–31	414–428
8	14.7	10–29	132–398	0.2	30–37	412–510
10	13.5	9–30	116–410	0.9	31–40	424–550
12	30.4	10–30	128–408	0.9	31–40	422–548
14	21.9	12–30	154–406	1.6	31–40	420–546
16	18.6	12–30	152–404	1.7	31–40	418–544
18	21.7	14–30	178–402	1.8	31–40	416–542
20	10.6	14–30	190–400	2.1	31–40	414–540
22	14.9	16–30	202–398	1.7	31–41	412–552
24	8.9	18–31	228–410	1.3	32–41	424–550
26	9.5	18–31	226–408	1.4	32–41	422–548
28				5.2	20–39	252–518
30				6.9	22–40	278–530
32				3.0	22–39	276–514
34				3.5	22–41	274–540
36				1.5	24–41	300–538
Total	175.4			33.7		

[a]Calculated from FI/MS analysis; *see* text.

The data in Table III classify approximately 55% of the aromatic hydrocarbons recovered by benzene as being asphaltene in nature. In contrast, Table XI leads to the conclusion that only approximately 16% of these hydrocarbons should be considered asphaltene. This latter value is in qualitative agreement with the prediction that only 8% of the hydrocarbons are asphaltene based upon the fact that 88% of the asphaltene neutrals are pentane soluble. The data in Table XI suggest, as intuitively expected, that the partitioning of aromatic hydrocarbons between compositionally unique oils and asphaltenes is a complex function of both the number of rings and the number of substituent alkyl carbons. However, a general tendency appears to exist for the minimum (maximum) number of alkyl carbons associated with an asphaltene (oil) aromatic hydrocarbon to decrease with increasing ring number.

The data in Tables V and VIII classify 59% of the volatile $C_nH_{2n+Z}O$ acids and 54% of the volatile $C_nH_{2n+Z}N$ bases as being asphaltene. In contrast, the weight percentages in Tables XII and XIII lead to the conclusion that only 13% of the volatile Z(O) acids and 12% of the volatile Z(N) bases belong to the asphaltenes. Furthermore, the same factors determining the aromatic hydrocarbons present in compositionally unique oils and asphaltenes appear to be generally applicable to the partitioning of these acids and bases.

Table XII. Summary Data for $C_nH_{2n+Z}O$ Aromatic Acids in Compositionally Corrected Oils and Asphaltenes[a]

	Oils			*Asphaltenes*		
		Range In			*Range In*	
−Z(O)	*Weight Percent × 10[1]*	*N*	*Molecular Weight*	*Weight Percent × 10[1]*	*N*	*Molecular Weight*
6	23.9	6–23	94–332			
8	14.8	9–24	134–344			
10	7.9	9–24	132–342			
12	11.9	10–24	144–340			
14	7.8	12–24	170–338			
16	9.5	12–24	168–336			
18	9.7	14–24	194–334			
20	5.7	15–24	206–332			
22	6.4	16–25	218–344			
24				3.8	17–26	230–356
26				4.1	18–26	242–354
28				3.3	20–27	268–366
30				2.3	21–29	280–392
32				1.2	22–27	292–362
34				0.2	25–27	332–360
Total	97.6			14.9		

[a]Calculated from FI/MS analysis; *see* text.

Table XIII. Summary Data for $C_nH_{2n+Z}N$ Aromatic Bases in Compositionally Corrected Oils and Asphaltenes[a]

	Oils			*Asphaltenes*		
		Range In			*Range In*	
−Z(N)	*Weight Percent* × *10*[1]	*N*	*Molecular Weight*	*Weight Percent* × *10*[1]	*N*	*Molecular Weight*
5	0.6	8–19	121–275			
7	0.8	8–18	119–259	0.1	19–20	273–287
9	1.0	8–20	117–285			
11	3.8	9–25	129–353			
13	2.0	11–24	155–337	0.1	25–28	351–393
15	1.5	12–25	167–349			
17	1.8	13–21	179–291	0.1	22–28	305–389
19	1.2	14–26	191–359	0.01	27–28	373–387
21	1.9	16–26	217–357	0.03	27–29	371–399
23	1.2	16–25	215–341	0.1	26–29	355–397
25	0.6	18–26	241–353	0.03	27–29	367–395
27				0.5	19–29	253–393
29				0.4	21–29	279–391
31				0.3	21–29	277–389
33				0.3	23–30	303–401
35				0.2	23–30	301–399
37				0.1	25–32	327–425
39				0.04	27–30	353–395
Total	16.4			2.3		

[a]Calculated from FI/MS analysis; *see* text.

Two simple examples suffice to seriously question the significance of the concept of oils and asphaltenes for the development of the science and technology of producing and upgrading coal liquids. First, significant interest exists in defining asphaltenes and oils in terms of general chemical structures and then relating these molecular structures to chemical/physical phenomena. However, the present analytical data for "as recovered" and "compositionally unique" fractions would result in significantly different conclusions concerning the average molecular weights and average molecular structures for oils and asphaltenes. However, in neither instance would these average structures correspond to molecules actually present. Furthermore, such average molecular structures may have little, if any, utility in factually analyzing or predicting the chemical/physical phenomena involved in the production and upgrading of coal liquids because of the well-known synergistic effects of structural entities (functional groups) on the physical chemistry of organic compounds (*58*, *59*).

Second, the chemical description and mathematical analysis of reaction networks require well-defined and compositionally unique products and reactants, and a knowledge of the stoichiometry and kinetic order for each reaction in the network. Clearly, the first requirement is not valid for oils and asphaltenes recovered by simple solvent extraction. Furthermore, the concept of oils and asphaltenes is inconsistent with the second requirement because the make-up of even compositionally well-defined fractions is determined by a set of unknown competing and consecutive chemical reactions of undetermined kinetic orders and stoichiometries. Thus, consideration of our detailed compositional results in terms of these fundamental chemical and mathematical principles clearly supports the contention of Neavel that to express reaction networks in terms of preasphaltenes, asphaltenes, and oils and to use rate laws based upon assumptions of the reaction order, though they may correlate specific experimental data, are unlikely to be fecund (*29*), at best and are likely to be wrong at worst.

In summary, the results from this and other studies (*30*, *31*, *32*) unambiguously demonstrate that simple solvent extraction of coal liquids does not yield chemically well-defined fractions. Consequently, detailed molecular analysis is a prerequisite for an in-depth analysis/prediction of the production and/or upgrading of coal liquids and for the correct evaluation of process effectiveness. In addition, implementation of routine process control/monitoring schemes employing fractions obtained from separation of products/reactants necessarily requires calibration by detailed molecular analysis. Finally, the separation method(s) should produce fractions possessing chemical significance.

Acknowledgments

We thank Dr. C. S. Hwang for assistance in developing computer programs for converting FI/MS ion intensities into quantitative distributions, Mr. G. Ritzhaupt for assistance in acquiring the IR spectra, and Dr. J. F. McKay for stimulating discussions of the IR spectra. Also, we thank the U.S. Department of Energy, contract numbers EX-76-C-01-2011, EX-76-S-01-2537, and ET-76-S-03-1811, for generous support of the research at Oklahoma State University.

Literature Cited

1. Bockrath, C. C.; Schweighardt, F. K. *Am. Chem. Soc., Div. Pet. Chem., Prepr.* (Washington, D.C., Sept., 1979) *24*(4), 949.
2. Weller, S.; Pelipetz, M. G.; Friedman, S. *Ind. Eng. Chem.* **1951,** *43*, 1572.
3. Liebenberg, B. J.; Potgleter, H. C. J. *Fuel* **1973,** *52*, 130.
4. Sternberg, H. W.; Raymond, R.; Schweighardt, F. K. *ACS Symp. Ser.* **1970,** *20*, 111.
5. Sternberg, H. W.; Raymond, R.; Aktar, S. *Am. Chem. Soc. Div. Pet. Chem., Prepr.* (Chicago, Aug., 1975) *20*(3), 711.
6. Yoshida, R.; Maekawa, Y.; Ishii, T.; Takeya, G. *Fuel* **1976,** *55*, 337.
7. Schwager, I.; Yen, T. F. *Am. Chem. Soc., Div. Fuel Chem., Prepr.* (San Francisco, Aug.–Sept., 1976) *21*(5), 199.
8. Schweighardt, F. K.; Sharkey, A. G. "Preprints, Coal Chemistry Workshop"; Stanford Res. Inst.: Stanford, 1976.

9. Farcasiu, M.; Mitchell, T. O.; Whitehurst, D. D. *Am. Chem. Soc., Div. Fuel Chem., Prepr.* (San Francisco, Aug.–Sept., 1976) *21*(7), 11.
10. Burk, E. H.; Kutta, H. W. "Preprints, Coal Chemistry Workshop"; Stanford Res. Inst.: Stanford, 1976.
11. Schweighardt, F. K.; Retcofsky, H. L.; Raymond, R. *Am. Chem. Soc., Div. Fuel Chem., Prepr.* (San Francisco, Aug.–Sept., 1976) *21*(7), 27.
12. Whitehurst, D. D.; Farcasiu, M.; Mitchell, T. O., EPRI AF-252, Project 410-1, 1976.
13. Schiller, J. E.; Farnum, B. W.; Sondreal, E. A. *Am. Chem. Soc., Div. Fuel Chem., Prepr.* (Chicago, Aug.–Sept., 1977) *22*(6), 33.
14. Schwager, I.; Lee, W. C.; Yen, T. F. *Anal. Chem.* **1977,** *49,* 2363.
15. Guin, J. A.; Tarrer, A. R., Department of Energy Report, FE 2454-6, Dist. Cat. UC-90d, 1977.
16. Bockrath, B. C.; Delle Donne, C. L.; Schweighardt, F. K. *Fuel* **1978,** *57,* 4.
17. Schwager, I.; Yen, T. F. *Fuel* **1978,** *57,* 100.
18. Thomas, M. G.; Granoff, B. *Fuel* **1978,** *57,* 122.
19. Tewari, K. C.; Kan, N.; Susco, D. M.; Li, N. C. *Anal. Chem.* **1979,** *51,* 182.
20. Schwager, I.; Yen, T. F. *Anal. Chem.* **1979,** *51,* 569.
21. Bockrath, B. C.; Schroeder, K. T.; Steffgen, F. W. *Anal. Chem.* **1979,** *51,* 1168.
22. Schwager, I.; Kwan, J. T.; Lee, W. C.; Meng, S.; Yen, T. F. *Anal. Chem.* **1979,** *59,* 1803.
23. Gary, J. H.; Golden, J. O.; Bain, R. L.; Dickerhoof, D. W., Department of Energy Report FE 2047-11, Dist. Cat. UC-90d, 1979.
24. Yen, T. F. *Am. Chem. Soc., Div. Pet. Chem., Prepr.* (Washington, D.C., Sept., 1979) *24*(4), 901.
25. Smith, P. A. S.; Romine, J. C.; Chou, S. P. *Am. Chem. Soc., Div. Pet. Chem., Prepr.* (Washington, D.C., Sept., 1979) *24*(4), 974.
26. Finseth, D.; Hough, M.; Queiser, J. A.; Retcofsky, H. L. *Am. Chem. Soc., Div. Petrol. Chem., Prepr.* (Washington, D.C., Sept., 1979) *24*(4), 979.
27. Tewari, K. C.; Li, N. C. *Am. Chem. Soc., Div. Pet. Chem., Prepr.* (Washington, D.C., Sept., 1979) *24*(4), 982.
28. Wen, C. Y.; Han, K. W., Department of Energy Report FE 2274-7, Dist. Cat. UC-90d, 1978.
29. Neavel, R. C. *Fuel* **1976,** *55,* 237.
30. Aczel, T.; Williams, R. B.; Pancirov, R. J.; Karchmer, J. H., Department of Energy Report FE-8007, Dist. Cat. UC-90d, 1977.
31. Aczel, T.; Williams, R. B.; Chamberlain, N. F.; Lumpkin, H. E. *Am. Chem. Soc., Div. Pet. Chem., Prepr.* (Washington, D.C., Sept., 1979) *24*(4), 955.
32. Ruberto, R. G. and Cronauer, D. C. *Fuel Process. Technol.* **1979,** *2,* 215.
33. Long, R. B. *Am. Chem. Soc., Div. Pet. Chem., Prepr.* (Washington, D.C., Sept., 1979) *24*(4), 891.
34. Steffgen, F. W.; Schroeder, K. T.; Bockrath, B. C. *Anal. Chem.* **1979,** *51,* 1164.
35. Benson, P. A.; Scheppele, S. E.; Greenwood, G. J.; Aczel, T.; Grindstaff, Q. G.; Beier, B. F., Department of Energy Report FE 2537-10, Dist. Cat. UC-90d, 1979.
36. FMC Corporation "Char Oil Energy Development," Res. and Dev. Report No. 56, Interim Rep. No. 1, Office of Coal Research, U.S. Department of the Interior, 1970.
37. Jewell, D. M.; Weber, J. H.; Bunger, J. W.; Plancher, H.; Latham, D. R. *Anal. Chem.* **1972,** *44,* 1391.
38. Ritzhaupt, G.; Devlin, J. P. *J. Phys. Chem.* **1977,** *81,* 521.
39. McKay, J. F.; Cogswell, T. G.; Weber, J. H.; Latham, D. R. *Fuel* **1975,** *54,* 50.
40. Snyder, L. R. *Anal. Chem.* **1969,** *41,* 314.
41. Petersen, J. C.; Barbour, R. V.; Dorrence, S. M.; Barbour, F. A.; Helm, R. V. *Anal. Chem.* **1971,** *43,* 1491.
42. Brown, F. R.; Friedman, S.; Makovsky, L. E.; Schweighardt, F. K. *Appl. Spectrosc.* **1977,** *31,* 241.
43. Conley, T. R. "Infrared Spectroscopy," 2nd ed.; Allyn and Bacon, Inc.: Boston, 1972.
44. Lumpkin, H. E.; Aczel, T. *Anal. Chem.* **1964,** *36,* 181.

45. Aczel, T.; Lumpkin, H. *Ann. Conf. Mass Spectrom. and Allied Topics, Atlanta, GA., 19th* 1971, 328.
46. Scheppele, S. E.; Grizzle, P. L.; Greenwood, G. J.; Marriott, T. D.; Perreira, N. B. *Anal. Chem.* **1976,** *48,* 2105.
47. Benson, P. A., M. S. Thesis, Oklahoma State Univ., 1979.
48. Scheppele, S. E.; Benson, P. A.; Greenwood, G. J.; Grindstaff, Q.; Aczel, T.; Beier, B. F. *Am. Chem. Soc., Div. Pet. Chem., Prepr.* (Washington, D.C., Sept., 1979) *24*(4), 963.
49. Scheppele, S. E., Department of Energy Report FE 2537-7, Dist. Cat. UC90-d; 1978.
50. Shultz, J. L.; Friedel, R. A.; Sharkey, A. G., Department of Interior, Bureau of Mines RI 7000, 1967.
51. Greenwood, G. J., Ph.D. Thesis, Oklahoma State Univ., 1977.
52. Scheppele, S. E.; Greenwood, G. J.; Pancirov, R. J.; Ashe, T. R. *Am. Chem. Soc., Div. Fuel Chem., Prepr.* (Houston, Mar., 1980) *25*(1), 37.
53. Boone, B.; Mitchum, R. K.; Scheppele, S. E. *Int. J. Mass Spectrom. Ion Phys.* **1970,** *5,* 21.
54. Ruberto, R. G.; Jewell, D. M.; Cronauer, D. C. *Fuel* **1978,** *57,* 575.
55. Snyder, L. R.; Buell, B. E. *Anal. Chem.* **1968,** *40,* 1295.
56. Farcasiu, M. *Fuel* **1977,** *56,* 9.
57. Farcasiu, M. *Fuel* **1978,** *57,* 576.
58. Leffler, J. E.; Grunwald, E. "Rates and Equilibria of Organic Reactions"; John Wiley & Sons: New York, 1963.
59. Hammett, L. P. "Physical Organic Chemistry," 2nd ed.; McGraw Hill: New York, 1970.

RECEIVED June 23, 1980.

6

Structure-Related Properties of Athabasca Asphaltenes and Resins as Indicated by Chromatographic Separation

M. L. SELUCKY,[1] S. S. KIM,[2] F. SKINNER,[3] and O. P. STRAUSZ

Hydrocarbon Research Center, Department of Chemistry, University of Alberta, Edmonton, Alberta, Canada T6C 2C2

Asphaltenes and resins from Athabasca bitumen were further separated by adsorption, ion exchange, and gel permeation chromatography. Molecular weights of asphaltene fractions from silica gel decreased with increasing fraction polarity while in ion exchange chromatography the highest molecular weight fractions were most strongly retained. GPC of more concentrated solutions of asphaltenes (≥ 1%) gave usual molecular weight distribution patterns while GPC of very dilute asphaltene solutions (0.01%–0.05%) showed time-dependent shifts toward low molecular weights. In all chromatographic methods, the overlap of asphaltenes and resins was considerable. The results were correlated with other experimental evidence (solubility, variation of yields with changes in precipitation conditions, and solubility of fractions obtained). It was concluded that definition of asphaltenes based only on solubility is not a satisfactory criterion and that the behavior of asphaltenes in chromatographic separations is incompatible with such structures where the polymer units are interconnected predominantly by σ-bonds. The asphaltenes are a complex state of aggregation best represented by the stacked cluster structure (micelle), which, however, cannot explain some of the GPC behavior of very dilute asphaltene solutions.

Recently, Bunger et al. (*1*) have pointed out, on the basis of simple thermodynamic reasoning, that substantial overlaps must exist between the maltene (i.e., deasphaltened bitumen) and asphaltene fractions with

[1]Current address: Alberta Research Council, Edmonton, Alberta, Canada T6G 2C2
[2]Current address: INHA University, Inchon, 160-01, Korea
[3]Current address: Raylo Chemicals, Edmonton, Alberta, Canada T6C 4A9

0065-2393/81/0195-0083$09.00/0

respect to molecular weight, and aromatic and heteroatom contents. Bunger also suggested earlier (*2*) that bitumen asphaltenes possess much lower molecular weights than measured by the common techniques and that the properties of asphaltenes are not inconsistent with such lower molecular weights (of the order of, e.g., 600–800 daltons). McKay et al. (*3*) share this opinion. By their very definition, asphaltenes are a chemically indeterminate mixture of compounds precipitated from bitumens, residua, or coals with nonpolar hydrocarbon solvents. This complex mixture contains compounds of a wide distribution of molecular weight [as shown by gel permeation chromatography (GPC)], polarities (as indicated by adsorption chromatography), and functionalities (as revealed by ion exchanger and adsorption chromatography, and IR spectroscopy). The term ion exchanger is preferred here since in aprotic solvents the mechanism involved is apparently not ion exchange. Similarly, the resins have been defined by a solubility criterion similar to the asphaltenes, that is, material soluble in *n*–pentane but insoluble in *n*-propane, and also by a chromatographic definition—that part of maltenes that cannot be eluted from clays with *n*-pentane.

Moreover, it has been shown (*4*) that a homologous series of *n*-alkanes, *i*-alkanes, and straight chain l-olefins furnish continuous precipitation curves; consequently the asphaltene precipitation approach using low molecular weight *n*-alkanes for the study of the higher molecular weight compounds present in the bitumens, residua, or coal derived liquids, seems to be lacking in the precision necessary for a basis for analytical studies. Since the precipitation curves flatten off starting with approximately nC_7 as precipitant, some authors prefer to use this solvent, thus increasing the proportion of resins and narrowing the range of compounds present in the precipitate. These authors will apparently get higher molecular weight values, lower H/C ratios, and higher heteroelements contents than those using nC_5 for precipitation. However, Koots and Speight (*5*) have shown that if asphaltenes and resins are separated from the Athabasca bitumen and then the asphaltenes are redissolved in the deresinated bitumen, or if the resins are replaced by the resins from another bitumen, any attempts at the complete redissolution of the asphaltenes fail. Furthermore, Boduszynski (*6*) precipitated nC_5 and nC_7 asphaltenes from two samples of the same material and obtained 17.0% and 10.6% precipitate, respectively, that is, nC_5 precipitated 6.4% more material than did nC_7. When the nC_5 asphaltenes were then redissolved in benzene and reprecipitated with nC_7, the yield was not 10.6%, as in the previous case, but was 14.8%. Similarly, *n*-decane precipitated only 5.9% asphaltenes, but when the nC_7 precipitate was treated with nC_{10}, the yield was 9.7%. Thus, the Speight results imply, and those of Boduszynski show, that the amount of precipitated asphaltenes (which is equivalent to the "quality" of asphaltenes) is affected by other components in this two-phase (solution/precipitate) system.

When part of a bitumen solution is precipitated, for example, by a nonpolar solvent, eventually an equilibrium or pseudoequilibrium will be established between the solution and the precipitate. Under such conditions, the Gibbs–Duhem equation should hold to a first approximation. It follows that the composition of the precipitate can be changed by introducing additional amounts of the solvent. In the same way, if the dissolved components are removed as in Boduszynsky's experiments described above, a compositional shift can again be expected, as shown by the increased amount of nC_7 asphaltenes obtained from the nC_5 asphaltenes, as compared with nC_7 asphaltenes obtained directly from the bitumen solution. This indicates clearly some of the problems that arise with an asphaltene definition based on solubility only. The solubility definition has been useful in such cases as fast characterization of feedstocks, since in natural products the various types of components will have a certain distribution around a median value.

However, we feel that disregard for the limitations of the solubility definition of asphaltenes is shown when man-made reaction mixtures (such as total products of asphaltene reactions) are treated so that the material after the reaction is dissolved in a minimum amount of benzene and precipitated with n-alkane, and the corresponding nC_5 or nC_7 precipitate is called "unreacted asphaltenes." Thus, thermolysis of asphaltenes will split off sidechains and other saturated moieties from the aromatic skeleton, and will eliminate some of the functional groups, thus removing heteroelements. The loss of sidechains will reduce the solubility of the material while the loss of functional groups will increase its solubility in nonpolar solvents. Lowering the molecular weight generally increases solubility, and aromatic condensation decreases solubility. Thus, in certain cases the "reaction asphaltenes" can be products of deep structural changes and not "unreacted asphaltenes." If such "asphaltenes" are not subjected to a more detailed analytical study in the same way as the low molecular weight products, much useful information escapes attention.

The second danger of this solubility definition of asphaltenes rests in the fact that once isolated as a "class," they may be—and often are—considered to be analogous to a compound and not recognized as a complex state of aggregation of countless individual species whose composition will change with the composition of other constituents of the original bitumen. Thus, Yen (*7*) has grouped asphaltenes with the multipolymers, that is, polycondensates with different repeating blocks; Ignasiak et al. (*8*) have included asphaltenes with the S-bridged polymers. We do not question the presence of such structural features as –S–,$-(CH_2)_n-$, or other types of bridges in the asphaltenes; however, the presence of such bridges may be a minor feature of the complex aggregate that constitutes the asphaltenes and may contribute only in a minor way to many of the properties of asphaltenes (*9*).

Some time ago work was initiated on chromatographic separations of asphaltenes and resins in various chromatographic systems (GPC, ion exchangers, adsorption) to compare their behavior in these analytical systems. Part of the results have been reported previously (*10*), but they are used here to provide a better overview of the results obtained from separating the same materials by various methods. Literature data on the various procedures were confined for the most part to one technique. These procedures were applied to different bitumens, thus precluding direct comparisons. See Discussion section for the relevant papers. The purpose of this chapter is to show that no separation method commonly used today permits a sharp differentiation between resins and asphaltenes. In these separations, considerable overlaps can be observed; the concept of increased stacking with increasing sample polarity is also open to question. Severe overlap of two fractions in a number of chromatographic separations is an indicator of their similarity in component distribution and overall composition. Thus, if two fractions overlap to a large extent, that is to say they have similar relative retention values in separations based on different properties of their components (such as size, polarity, functionality, as brought about by GPC, adsorption and ion exchanger chromatography, respectively), they can be expected to be very much alike. We will show that with a material such as Athabasca bitumen, these overlaps are so extensive that the compounds present are obviously parts of a continuum of structures; furthermore, asphaltenes do not represent fixed but rather loose, concentration-dependent aggregates. It is the latter property that makes the study and understanding of these bitumen (residua or coal liquid) fractions so difficult.

Experimental

The adsorption chromatography on silica gel and clays, ion exchanger chromatography, and gel permeation chromatography were carried out using conventional procedures. Experimental details are described in the following series of references: precipitation of nC_5 asphaltenes (*11, 12*); separation of resins (*13, 14, 15*); ion exchanger chromatography of asphaltenes (3), and of maltenes (*11, 12*).

The gel permeation chromatography was done using either 2 cm × 120 cm steel columns packed with Styragel 1000 Å (37–75μ) for preparative separations [and for analytical separations 30-cm μ-Styragel 100 Å + 30-cm μ-Styragel 500 Å (3/8")] or μ-Styragel 1000 Å (30 cm × 3/8") columns (all from Waters Assoc.) and Waters Assoc. HPLC, Model 202 with RI and UV (254 nm) detectors. The eluent was methylene chloride in all cases. All chromatographic solvents were of reagent grade, dried and fractionated before use.

Elemental analyses and molecular weight determinations (VPO) were done by K. Bernhardt Laboratories, Engelskirchen, West Germany. The IR analysis was made with a Nicolet FT-interferometer from chloroform or methylene chloride casts in the absorbance mode.

Results and Discussion

In the laboratory, the standard procedure for obtaining the main three fractions of a bitumen consists of separating the asphaltenes by solvent precipitation; then removing the resins from the deasphalted sample by chromatography on clays (Attapulgus clay, Florisil); and finally separating the deasphaltened, deresinated sample, conventionally called the oil fraction, by chromatography on silica gel and alumina to obtain the various classes of hydrocarbons. Recently, ion exchangers have also been used for the separation of acids and bases from the deasphaltened sample (*16*). The latter method furnishes fractions of acids and bases, but as might be expected, the gravimetric total results do not agree between the different separation methods (*17*). The difference amounted to 6%–8% with the maltene fraction of Athabasca bitumen and also with the maltene fraction of Cold Lake bitumen. Thus, the differential amount between the various methods of separation ended in the polyaromatic fraction of the USBM-API 60 procedure, which gave consistently lower amounts of total combined acids, bases, and Lewis bases (*11, 12*) than the yield of resins in separations on clay. Whereas the original USBM-API 60 procedure was devised for distilled fractions plus residuum, we have shortened the analysis by omitting the tedious distillation step (*11, 12*); we have studied in detail the adverse effects of these short cuts. Later, Bunger et al. (*18*) did comparative analyses of various bitumens, omitting the preseparation of asphaltenes by precipitation. Their results suggest that all of the asphaltenes present appeared as Lewis bases. However, we found, in repeated experiments, only 20% of the precipitated asphaltenes passed both ion exchangers (*10*). This discrepancy is so profound that we decided to study the behavior of asphaltenes and resins from the same type of bitumen in the various chromatographic systems. Our goals were to shed some light on the scope and limitations of the various separation methods and to obtain a better understanding of the fate of the nonoily bitumen fractions in these separations.

Chromatography of Resins and Asphaltenes on Silica Gel. Woelm silica gel activated at 140° C for 4 h is a stronger adsorbent than activated Attapulgus clay (*19*); therefore, on elution of the maltene fraction of Athabasca bitumen on silica with nC_5, we found that this eluent is not a strong enough solvent to separate the oils from the resins in a manner that agrees quantitatively with the separation on clay. Various proportions of benzene, from 15%–10% in the nC_5 eluent, showed that with a 15% Bz/nC_5 eluent (following elution with nC_5), a fraction was obtained that upon subsequent separation on alumina (Woelm 200N, activity super I) gave a benzene eluate amounting to 24% of the bitumen (*14*). Since we know from previous work that the bulk of this benzene fraction consists of tri- and higher aromatic hydrocarbons, the 15% Bz/nC_5 solvent was apparently slightly too strong. In terms of resins transferred into the hydrocarbon fraction this represents 2.6% of the bitumen. By trial and error, 12% Bz/nC_5 was found to be the best compromise to obtain yields of resin comparable with those obtained with clay. It is not implied here that the composition of the hydrocarbon fraction from silica gel was strictly identical with that from the clay, but with the heavy oil and bitumens of Northern Alberta the bulk results compared well. The combined resin eluates from silica gel represented about 25% bitumen when 12% Bz/nC_5 was followed by 50% Bz/nC_5. The remainder of the resins (13.4% bitumen) was then removed with tetrahydrofuran or benzene/methanol/ethyl ether.

Asphaltenes could not be applied on silica gel in the same manner as the resins because of their insolubility in nC_5. Therefore, the sample of asphaltenes was dissolved in benzene, and this solution was stirred with sufficient amount of silica gel at 50°C in a rotary evaporator. Evaporation of benzene in vacuo furnished asphaltene-coated silica gel, which was added to the top of the silica gel column. The weight ratio of asphaltene to total silica gel was 1:100. The column was first eluted with nC_5, then with benzene/nC_5 mixtures of increasing proportions of benzene up to 100% benzene. Table I shows that already nC_5 and 30% Bz/nC_5 eluted 7.5% of the asphaltenes, 45% Bz/nC_5 another 3.5%, and 60% Bz/nC_5 an additional 6.6%; altogether 17.6% asphaltenes were eluted with this series of solvents. Under very similar conditions, the elution of resins amounted to 60% of total resins.

Table I. Separation of Athabasca Asphaltenes on Silica Gel

Fraction	*Weight Percent*	*Cumulative Weight Percent*	*Molecular Weight*
n-Pentane	0.3		—
nC_5/15% Bz[a]			—
nC_5/30% Bz-I	5.4		8200
nC_5/30% Bz-II	1.9	7.5	6400
nC_5/45% Bz	3.5	11.0	4900
nC_5/60% Bz-I	3.8	14.8	4400
nC_5/60% Bz-II	2.8		5400
nC_5/80% Bz	3.4	21.0	3500
Bz	2.5	23.5	3500
Bz/15% CH_2Cl_2	2.5	26.0	2600
Bz/50% CH_2Cl_2	1.1		4200
CH_2Cl_2	1.7		2900
CH_2Cl_2/15% MEK[b]	17.3		—[e]
CH_2Cl_2/50% MEK	1.1		—
MEK	1.5		—
MEK followed by Bz	0.9		—
MEK followed by THF[c]	22.9	72.6	—
THF/15% MeOH-I	4.5		—
THF/15% MeOH-II	12.0	89.0	—
THF/50% MeOH	2.5		—
followed by Bz	0.5		—
followed by CH_2Cl_2	—		—
followed by THF	0.6		—
THF/15% H_2O	6.6		—
THF/10% *i*PrN[d]	0.5		—
THF/25% *i*PrN	0.9	100.6	—

[a]Bz = benzene.
[b]MEK = methyl ethyl ketone.
[c]THF = tetrahydrofuran.
[d]*i*PrN = isopropyl amine.
[e]Molecular weights were not determined, either for lack of sample or because of some problems with THF and MEK.

A study of Table I reveals several interesting facts. Firstly, the first fraction obtained in a sufficient amount for vapor pressure osmometer (VPO) measurements (30% Bz/nC_5-I) had the highest molecular weight (MW) of all asphaltene fractions eluted from silica gel measured so far (8200 daltons). Secondly, material eluted with this composite solvent was travelling through the column as two discrete bands. Thirdly, this highest MW material gave an IR spectrum that most closely approached the spectrum of a high MW pure hydrocarbon fraction, with little contribution from functional groups, as indicated by IR analysis (*see* Figure 1).

Further, progressively eluted fractions showed certain trends of MWs. Thus, the MW dropped from 8200 to 4400. As the polarity of the solvent was progressively increased, a second series of compounds was eluted whose MWs started at 5400 daltons and ended with 2600 daltons; the last two fractions obtained in sufficient amount for VPO measurements again had MWs that declined from 4200 to 2900 daltons. No oxygen-containing solvent had been used so far, so that MW artifacts are improbable. It seems that the asphaltenes are eluted as a number of discrete series of material of increasing polarity and declining MW. It is immaterial in this context if "size" effects are actually overlapping polarity effects. The IR spectra clearly show that the polarity of fractions was indeed increasing with elution volume. This trend of MWs was unexpected, as it is generally thought that functional groups aid associations through possible hydrogen bonding, charge-transfer complexes, or charge separations, and, therefore, the MW would increase with increasing functionality. Also, in these separations the ratio of 1380 cm^{-1} to 1450 cm^{-1} peaks remained approximately constant, while in general, the peaks in the 1700–1800 cm^{-1} region were increasing with noticeable shifts of maxima between the fractions. The same could essentially be said about the hydrogen bonding and the 1300–1000 cm^{-1} regions. A more detailed study of the IR spectra, which is not the subject of the present paper, will reveal that, frequently, distinct types of material are being concentrated and separated by elution chromatography.

Thus, IR has confirmed that, whatever the elution mechanism is, the "polarity" of the first fractions (measured by the relative size of peaks at 3500, 1700, 1600, and 1250–1000 cm^{-1}) is the lowest of all fractions collected. Therefore, postulates that increasing polarity tends to contribute to MW may require revision.

The first three solvent systems eluted from silica over 17% of the applied asphaltenes while the quantity of neutral asphaltenes from ion exchanger chromatography, that is, those components that passed both the anion and cation exchanger columns, amounted to 20% of the material. This is a reasonably close agreement between fractions of asphaltene separated by two different methods. At first sight, this suggests that these fractions are very similar; however, comparison of MWs of these fractions does not support this conclusion.

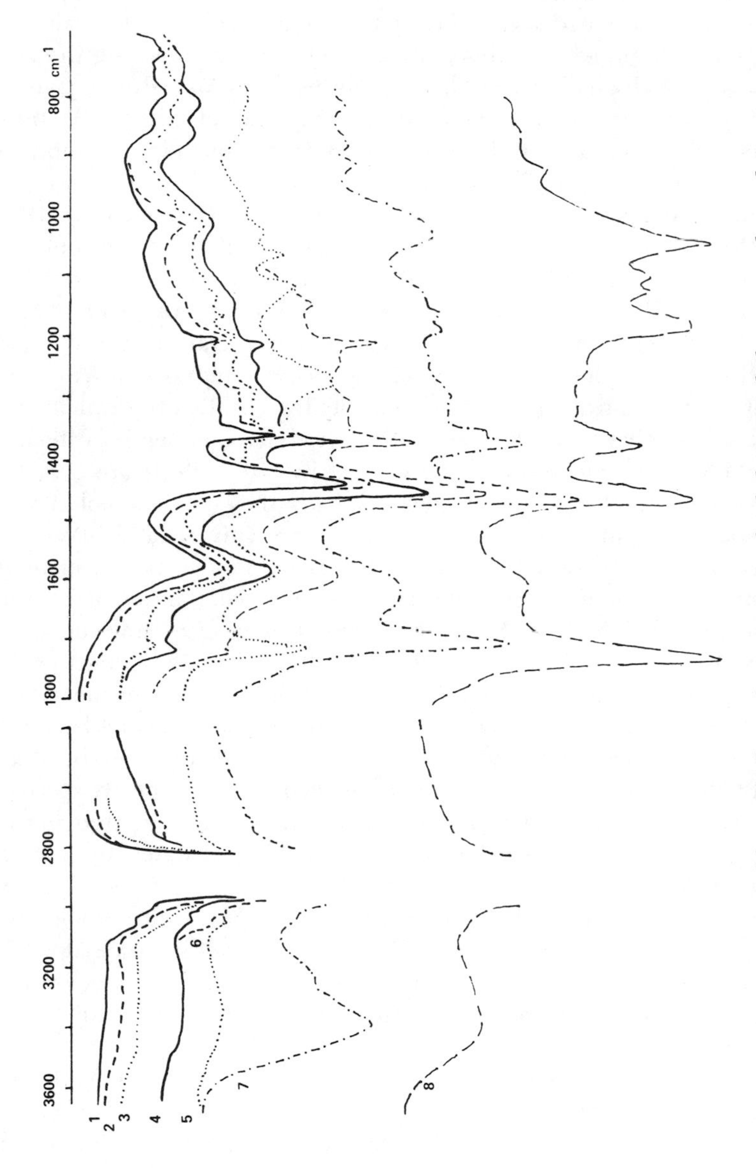

Figure 1. FT-IR spectra of fractions from the separation of Athabasca asphaltenes on silica gel: (1) 30% Bz-I (Cyclohexane was used instead of n-*pentane); (2) 30% Bz-II; (3) 60% Bz-I; (4) 80% Bz; (5) 15% CH_2Cl_2/Bz; (6) 50% CH_2Cl_2/Bz; (7) 100% MEK; (8) MEK/Bz; (Bz = benzene, MEK = methyl ethyl ketone)*

Also a present belief is that a large proportion of asphaltenes would be chemisorbed on the silica gel; this was not the case. It was evident from column color that only a very small amount of the asphaltene sample was retained. Comparison of gravimetric results showed that desorption in the neighborhood of 100% was achieved. It should be noted here that some of the solvents used dissolved minute amounts of silica. (Experiments with alumina have shown that alumina is also slightly soluble in some solvents). Extreme care was taken to remove these traces of silica from the isolated fractions.

Thus, these experiments can be summarized by stating that there is an apparent, considerable overlap of retention volumes of various compound classes in the separation of resins and asphaltenes on silica gel at sample to gel ratios of 1:100. Also for asphaltenes the general trend of molecular weights was towards lower values with increasing sample polarity, as evidenced by IR spectra. Absorption in the hydrogen-bonding region was completely absent in the highest MW fraction (nC_5/30% Bz–I) and also, in the 1800–1700 cm^{-1} region, only a trace absorption could be noticed. The separations of sufficient amounts of asphaltenes require large columns (to keep column loading low) and correspondingly large volumes of solvents, making the separation time-consuming.

Separations of Asphaltenes and Resins on Ion Exchangers. Ion exchangers have been introduced in the so-called USBM-API 60 procedure (*16*) for the separation of acids and bases from distillation fractions, and by us (*11, 12*) for the separation of acids and bases from fractions separated by precipitation (asphaltenes) and by chromatography, without prior distillation. Later, McKay et al. (*3*) applied essentially the same procedure to the separation of Wilmington asphaltenes, and Bunger et al. (*18*) to the separation of whole bitumens.

For the separation of asphaltenes, nC_5 had to be replaced by cyclohexane (*3*) for solubility reasons. We have used the McKay procedure, later somewhat modified, because the samples of Athabasca asphaltenes were not completely soluble in cyclohexane. The separated asphaltenes (nC_5) were applied on both possible column sequences (i.e., anion/cation exchanger and cation/anion exchanger). The gravimetric results are summarized in Figure 2 (*10*). While the amount of material passing through both columns with cyclohexane eluent is the same within experimental error, irrespective of the sequence of ion exchangers, the relative proportions of the retained fractions differ for the two ion exchanger sequences. Thus, with the anion exchanger first, the percentage of acids was 56.6% and that of the bases only 23.3%, while by reversing the ion exchanger sequence, the yield of bases increased to 39.6% and that of the acids fell to 39.8%, that is, 16% of the asphaltenes depended in their elution upon the ion exchanger sequence. If it is really only acids that interact with the anion exchanger and bases that interact with the cation exchanger, and this is not known with any degree of certainty, the results would suggest that 16% of the material has an amphoteric character in addition to about 20% neutral material and about 40% acids and 23% bases. This is in sharp contrast with the results of McKay obtained on Wilmington asphaltenes (3) and also with the results of Bunger (*18*), who separated Athabasca bitumen without precipitation of asphaltenes. In the latter reference, it seems from gravimetric results that all of the asphaltenes appeared as Lewis bases, that is, material retained on Fe^{3+}/IRA-904 column. It is difficult to explain these differences. It is possible that Wilmington asphaltenes do differ greatly from Athabasca asphaltenes. On the other hand, the discrepancy between our results and Bunger's results cannot be explained so simply. That the presence of resins in the latter work influenced the behavior of asphaltenes to such an extent does not seem probable. Since relatively concentrated solutions of asphaltenes were used in preparative work and coarse ion exchanger particles, the mass transfer must be relatively slow. Therefore, it was attempted to demonstrate that relatively fast sample application led to the appearance of the bulk of material in the cyclohexane eluent and that there existed an optimum time of sample contact with the ion exchanger resin prior to elution. It was found that the optimum time of sample application was, in this case, 1–2 h. At application times of ½ h or less, the bulk of nonretained material increased, and at application times exceeding 4 h, more sample remained irreversibly retained. Table II shows three experiments where the first two—run by different operators—were carried out under identical conditions while the third sample was allowed to interact 4 h with the resin before solvent circulation was started.

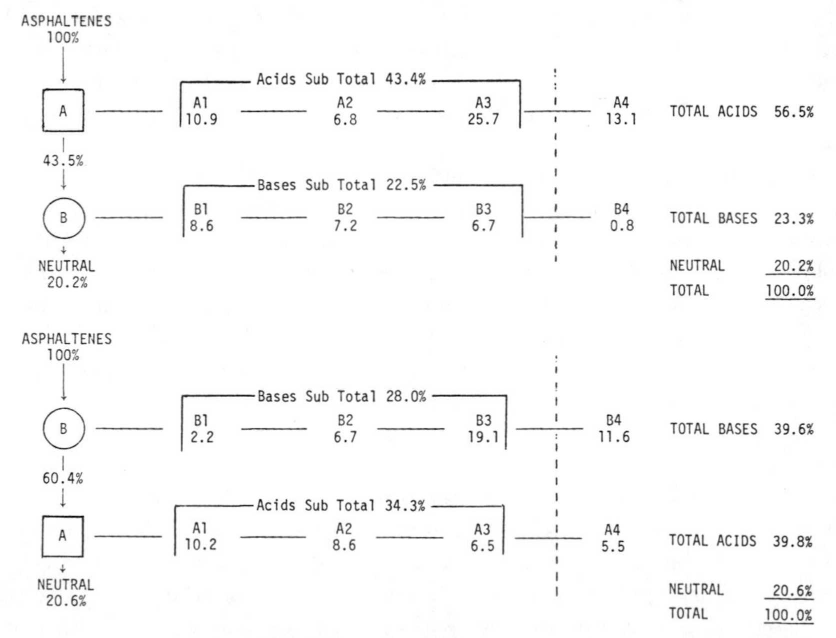

Figure 2. Separation of Athabasca asphaltenes on ion exchangers: A = acids (separated on A-27); B = bases (separated on A-15); vertical dashed line represents the end of solvent system described in (3)

Table II. Separation of Athabasca Asphaltenes on Anion Exchange Resin IRA-904

	Wt. %		
Fraction	*Operator Number 1*	*Operator Number 2*	*Operator Number 3*[a]
Bz	50.2	52.9	49.5
Bz/MeOH	6.0	5.4	3.6
Bz/MeOH/CO_2	14.5	15.4	11.1
CH_2Cl_2/CO_2	23.1	22.4	22.7
Bz/AcOH	4.4	4.2	4.4
Bz (60 h)			2.0
Total	98.2	100.3	93.3

[a]Asphaltenes allowed to interact with exchange resin 4 h before solvent elution started.

Two additional problems occurred in these separations. Firstly, the asphaltene samples when applied from cyclohexane solutions as described by McKay (*3*) were leaving 3%–4% residuum at the column head. Secondly, the solvent series was the same as that used by McKay, differing from that used for resin separations only in that cyclohexane was used as the initial solvent instead of nC_5, and benzene/methanol was employed instead of methanol. These modifications are necessary because of the lower solubility of asphaltenes.

McKay's solvent sequence completely eluted the Wilmington asphaltenes but did not elute all the Athabasca asphaltene samples and had to be extended by additional solvent mixtures to obtain good sample recoveries (cf. Figure 2). For large scale preparative separations of asphaltenes, the asphaltenes were dissolved in benzene and eluted with the same solvent, omitting the cyclohexane step. This accelerated the operation, but at the same time, as expected, the percentage of the neutral fraction now increased from 20%–21% to approximately 28%–30%, in reasonable agreement with the bulk results from the cyclohexane experiments (*see* Table III). Table III also shows the additional solvent systems used.

The separation of resin acids and bases has been described previously (*11*, *12*), *see* also comments in (*14*). The percentage of resins, as determined for deasphaltened Athabasca bitumen by their separation on an Attapulgus clay column, was 34%, while combined acids, bases, and Lewis bases amounted to 25.9% of the whole bitumen. Thus, 8.1% of material retained as resins on Attapulgus clay did not interact with the ion exchangers or the complexation column and appeared in the polyaromatic fraction. The distribution of material within the resin fraction was 46.1% acids, 21.9% bases, and 32% neutral compounds. Thus, the pattern of acid and base distribution is similar for the resins and asphaltenes, except for a higher proportion of neutral material present in the resins.

Table III. Preparative Scale Separation of Athabasca Asphaltenes on Ion Exchangers IRA-904 and A-15

Fraction			*Weight Percent*
	Asphaltene Acids		
904 Anion Exchange Resin			
Bz	50.2		
Bz/MeOH	6.1	acids	6.1[b]
Bz/MeOH/CO_2	14.5	acids	14.5
$CH_2Cl_2CO_2$	23.1	acids	23.1
Bz/AcOH	4.4	acids	4.4
	98.3		
	Asphaltene Bases		
A-15 Cationic Exchange Resin			
Bz	58.3[a]	neutral	29.3[b]
Bz/MeOH	5.9	bases	3.0
Bz/MeOH/*i*PrN	14.1	bases	7.1
Et_2O/*i*PrN	0.6	bases	0.3
THF/*i*PrN	15.6	bases	7.9
Percent Recovery	94.5		
Total acids + neutral + bases			95.7

[a]Percent of material applied.
[b]Percent of the asphaltene fraction.

As mentioned above, for the preparative separation of asphaltenes into fractions on ion exchangers, using the sequence of solvents suggested by McKay et al. (3) left 27.5% of the applied sample on the anion exchanger column and 16.2% of applied sample on the cation exchanger column. More complete elution could be achieved by adding methylene chloride/carbon dioxide and benzene/acetic acid eluents to the anion exchanger solvent sequence and ethyl ether/*i*-propylamine and tetrahydrofuran/*i*-propylamine to the cation exchanger eluents. Table III shows the amounts of material in these additional fractions. Omission of cyclohexane as the first solvent led to an increase in the nonretained fraction from either ion exchange resin. The initial concept was that if needed, this fraction could be rechromatographed using cyclohexane. However, once evaporated to dryness, the benzene eluate from the anion exchanger possessed very little solubility in cyclohexane and was only partially soluble in benzene, even after prolonged agitation in a rotary evaporator at 60° C. This suggests that upon withdrawing part of the material from the bulk of the asphaltene sample certain rearrangement takes place, resulting in a material whose properties (after solvent evaporation) differ from those of the initial asphaltenes in solution. It should not be overlooked that the asphaltenes were isolated as benzene solution. Therefore, there was little likelihood that solvent effects could be responsible for the loss in solubility; the same may be said of the treatment at mild temperatures (60° C, nitrogen blanket).

In contrast to the results from sorbent elution (silica gel), the neutral fraction from ion exchangers had rather low MW (Table IV), 4900 daltons, which became still lower in separations where cyclohexane was omitted from the sequence (Table V). However, the latter data are not strictly comparable since these analyses were for comparisons with GPC data; therefore, methylene chloride was used as solvent in Table V.

The IR spectra of asphaltene fractions from the IRA-904 anion exchanger are shown in Figure 3. Again, it may be noticed that the cyclohexane fraction (even though this fraction from the anion exchanger still contains some bases) is the most neutral fraction of all, void of absorptions in the hydrogen-bonding and —CO— regions and with the smallest background in the 1300–1000 cm^{-1} region. On the other hand, ^{19}F spectra of trifluoroacylated samples of these asphaltene fractions showed the presence of acylable groups (*20*). The precipitated asphaltenes contained 0.4 OH (acylable groups expressed as OH groups) per kg of material. The benzene/methanol fraction, which by analogy with the separation of resins should be a phenolic concentrate, contained only 0.1 OH/kg while the benzene/methanol/carbon dioxide fraction, which should represent stronger acids, contained 1.8 OH/kg. The last two fractions, eluted with methylene chloride/carbon dioxide and benzene/acetic acid, contained only 0.3 and 0.1 OH/kg, respectively. However, these data are not very reliable because solubility problems have been encountered, especially with fractions benzene/methanol/carbon dioxide and methylene chloride/carbon dioxide, where only about 25%–30% dissolved in the NMR solvent (deuteromethylene chloride). This was a rather general problem, as often the separated fractions were much less soluble in organic solvents than were the total asphaltenes, or a fraction eluted with a certain solvent was reluctant to redissolve in the same solvent once the solvent was evaporated and newly added. Also, MW determinations of all the fractions have shown that the most strongly retained fractions (both from the anion and cation exchanger) had the highest MW of all the separated fractions.

Similar solubility phenomena can also be observed for the resins. The sample for resin separation after removal of asphaltenes is normally applied as a solution in nC_5. However, after the removal of the oil, the fractions obtained from the ion exchangers show a marked decline in solubility in the same solvent. Also, if resin separation is done by the SARA method, the tetrahydrofuran fraction is hardly soluble in nC_5. This again shows the solubility criterion to be a function of several variables; the removal of some of the solubilizing components of the resins renders the remainder insoluble.

Moreover, these results strongly suggest that, generally, once the material (asphaltenes or resins, or even the whole bitumen) is separated into various fractions, or to be more explicit, once part of the material is removed from the bulk, certain rearrangements of the associated structures may take place, leading to more strongly agglomerated entities. This phenomena is, in our opinion, also an indication that smaller entities associate into larger ones such as in the micelle model of Dickie and Yen (*21*). Thus, experimental evidence

Table IV. Analyses and Element Ratios of Athabasca Asphaltene Fractions from Ion Exchangers

Fraction	*C*	*H*	*N*	*O*	*S*	*MW*	*H/C*	*N/C × 100*	*O/C × 100*	*S/C × 100*	*Percent of Asphaltene*
Acids:											
A-27 Anion Exchanger											
Cyclo-C_6	80.9	8.1	1.2	2.5	7.4	4900	1.20	1.27	2.32	3.43	43.5
Bz	80.9	8.0	1.2	2.0	7.9	5300	1.19	1.27	1.85	3.65	10.9
Bz/MeOH	78.4	8.1	1.2	4.8	7.6	3400	1.24	1.31	4.59	3.54	6.8
Bz/MeOH/CO_2	76.9[a]	9.2[a]	0[a]	11.2[a]	2.6[a]	500[a]	1.44[a]	—	10.92[a]	1.27[a]	25.7
Bz/AcOH	77.1	8.0	1.2	6.8	6.9	4800	1.25	1.33	6.61	3.36	13.1
Bases:											
A-15 Cation Exchanger											
Cyclo-C_6	79.5	8.0	1.1	2.8	7.8	4900	1.21	1.19	2.64	3.58	20.2
Bz	60.7	7.8	1.1	2.8	7.9	3800	1.16	1.17	2.50	3.67	8.6
Bz/MeOH	78.7	8.0	1.0	3.7	8.6	2400	1.24	1.09	3.53	4.10	7.2
Bz/MeOH/*i*PrN	77.2	8.2	2.1	5.7	6.7	2700	1.27	2.33	5.54	3.25	6.7
THF/*i*PrN	79.0	8.5	1.2	3.2	8.2	5000	1.29	1.30	3.04	3.89	0.8
Asphaltene sample	80.0	8.2	1.2	3.2	8.2	4550	1.23	1.29	2.63	3.70	

[a] Sample contained benzoic acid (from the exchange resin).

Table V. Analyses and Element Ratios of Athabasca Asphaltenes (Cyclohexane Omitted from Elution Sequence)

	Percent C	*Percent H*	*Percent N*	*Percent O*	*Percent S*	*Molecular Weight*[a]	*Percent Ash*		*Percent Asphaltene*
ASP	77.98	8.02	0.99	2.74	7.88	4000	2.12	light brown	
IRA-904									
Bz	78.90	7.95	0.90	2.64	7.56	2200	2.03	light brown	56.1
Bz/MeOH	79.40	9.33	0.78	5.29	4.29	1450	0.70		5.3
Bz/MeOH/CO_2	78.96	8.44	0.81	4.76	7.04	2700	0.16		11.3
CH_2Cl_2/CO_2	80.18	7.91	1.13	2.85	7.86	4700	0.06		21.2
Bz/AcOH	75.24	7.77	0.95	6.94	7.05	5500	2.13	light brown	6.0
A-15									
Bz	76.97	8.02	0.74	3.19	7.67	1750	3.47	light brown	29.5
Bz/MeOH	77.69	8.44	0.93	3.84	8.85	2200	0.42		7.8
Bz/MeOH/*i*PN	78.83	8.15	1.45	3.24	7.42	3300	0.95		8.9
THF/*i*PN	80.80	7.82	1.27	1.75	8.23	6500	0.06		9.7

[a]Solvent: CH_2Cl_2.

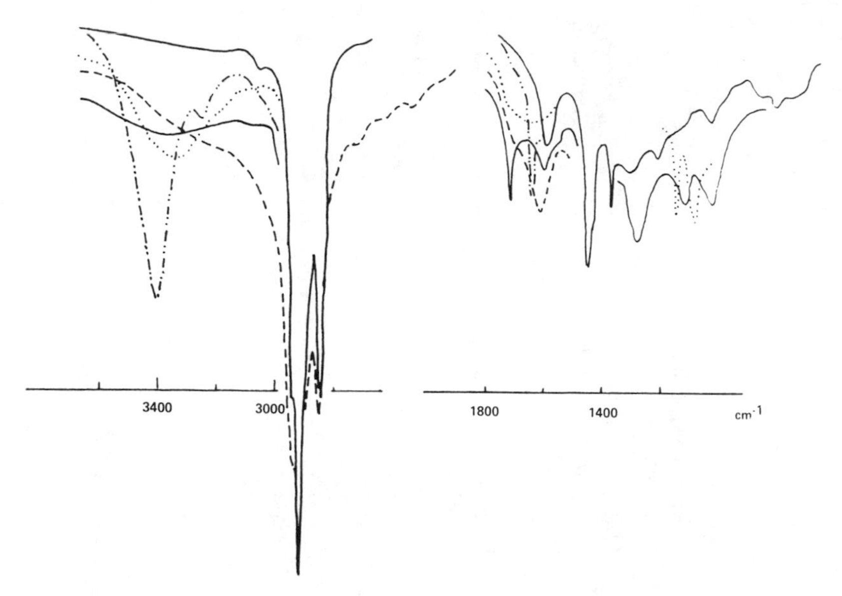

Figure 3. FT–IR spectra of Athabasca asphaltene acids fractionated on A-27: (——) upper curve, cyclohexane; (——) lower curve, Bz; (. . . . .) Bz/MeOH; (- - - -) Bz/MeOH/CO_2; (- · · · · · · -) Bz/AcOH fraction

gets slowly accumulated to the effect that the whole system of higher MW compounds in a bitumen is a dynamic system where association of various entities results in the formation of different agglomerates, and where the final character of the agglomerate depends to a large extent on which entities have been removed from the whole.

Thus, MW determinations, to some extent, lose their significance. If, for example, we take all the isolated fractions from the ion exchangers (about 98% recovery) and recalculate the MW of the original asphaltenes, we obtain the following: the contribution of 92% of the material would additively lead to a MW of 2750, that is, the residual 8% would have to have a MW of about 15,700 to make the total asphaltenes reach the experimental value of MW 4000. This cannot be the case, as will be seen from gel permeation measurements that were done with the intention of shedding at least some light on the discrepancies observed so far. A more detailed discussion on MW problems will follow the section on GPC.

Gel Permeation Chromatography. More and more work with residua and asphaltenes is using GPC as a useful separation method for size separations of the heaviest fractions of bitumens and petroleum residua. In a manner similar to that used in polymer chemistry, attempts have been made to calibrate a GPC MW scale so that it may be brought into accord with MW measurements by other methods (e.g., VPO). In other words, we are trying to relate the peak value (V_a) and statistical distribution of the MW across the

GPC peak to the number-average MW of the material as measured by VPO. This relation also helps in another way. The GPC band visually shows the statistical distribution of MWs over the whole profile, provided that no sorption effects contribute to the skewness (tailing) of the GPC band. Practice has shown that this is most probably not the case, and in carefully executed experiments the tailing of the peak is clearly visible. Reerink and Lizjenga (*22*) have pointed out that this tailing may adversely influence the quality of the results. For good results, correction should be made for the differences in detector response over the MW profile. They also pointed out two additional problems: first, that the universal Benoit equation relating the product of MW and limiting viscosity (M_η) to elution volume [V_R] does not hold for asphaltenes. As monodisperse polystyrene samples have much larger hydrodynamic volumes than asphaltenes, they cannot be used for straightforward calibration purposes. Previous work used ultracentrifuge measurements for determining the MW of fractions and plotted them against the differential (cumulative weight) of the fraction that represents the first derivative of the distribution function. As the curve for ultracentrifuge measurements is very similar to that showing cumulative weight fraction against V_R, the results from one were transferred onto the other curve by rectagonal operations. The low molecular weight part, where ultracentrifuge separations cannot be done, was calibrated using MWs from VPO. The calibration is rather involved as MW from VPO does not tally with the GPC peak maximum, and also the final step in the calibration uses iterative calculations to determine coefficients of an empirical equation. The same authors have correctly criticized previous work of Snyder and Altgelt (*23*, *24*), as, for example, Altgelt's fractions might have been polydisperse and, most probably, the same held for the fractions investigated by Snyder. From the data of Reerink and Lizjenga (*22*), best exemplified by the plot of log (MW) against V_R for polystyrenes and asphaltenes, they argue that the asphaltene curve should be above the polystyrene curve, which also implies that asphaltenes should be eluted later. They actually obtained, at equal V_R, MW ratio M_A/M_{PS} (A = asphaltenes, PS = polystyrene) of 1.5–2 for the various samples investigated.

In Figure 4 the results of this work are plotted. The asphaltene curve had a steeper slope than that for polystyrenes, in accordance with the results of Reerink and Lizjenga (*22*), but most of the straight part lay under the curve of polystyrene. Also, with the column used—Styragel 1000 Å—the asphaltenes started to become excluded at lower MWs than the polystyrenes. The discrepancy most probably lies in the difference in eluent or in the method used for MW measurement, or both. We separated a number of GPC fractions on Styragel 1000 and determined the number-average MWs by VPO. Since we needed to make a number of comparative measurements only, the accuracy of MW values was not of primary importance. It should also be pointed out that regression analysis has shown that the points for dodecylbenzene and toluene also lay on the polystyrene calibration curve (*see* Figure 4).

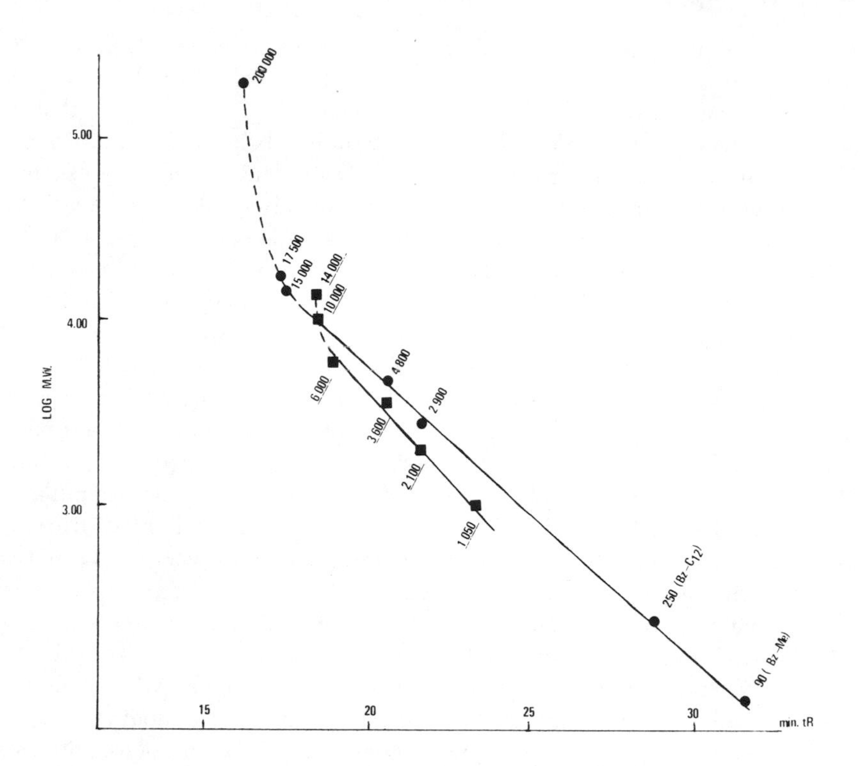

Figure 4. Calibration of preparative Styragel 1000-Å column: (●) = polystyrene standards; (■) = asphaltene fractions; MWs of asphaltene fractions determined by VPO from benzene

A study of MW distribution for precipitated asphaltenes and the derivation of conclusions about bitumen or asphalt properties from it has severe limitations since this complex mixture exhibits a considerable overlap of GPC curves for all the fractions obtained in a conventional separation procedure. Similarly, the resins separated on clay and the eluted hydrocarbons exhibit overlap, as shown by Figures 5 and 6. Figure 5 demonstrates the GPC profiles of Athabasca asphaltenes (nC_5) and resins (Attapulgus clay—total resin eluent)

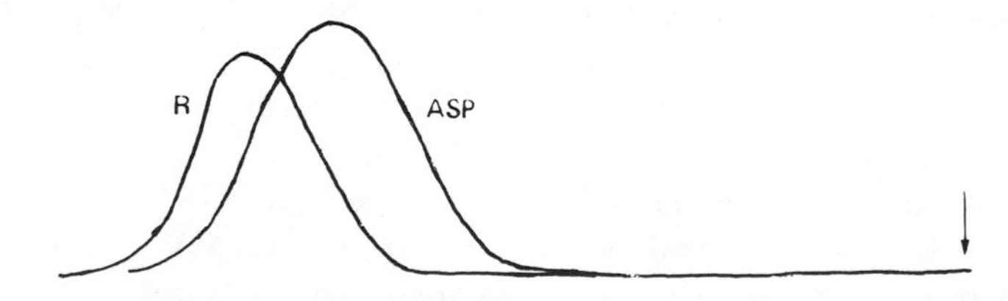

Figure 5. Preparative GPC separation of Athabasca asphaltenes (ASP) and resins (R): column, 120 cm × 2 cm i.d., Styragel 1000 Å (37–65 μ); RI detector × 16; CH_2Cl_2 at 9 mL/min; CS = 0.2"/min

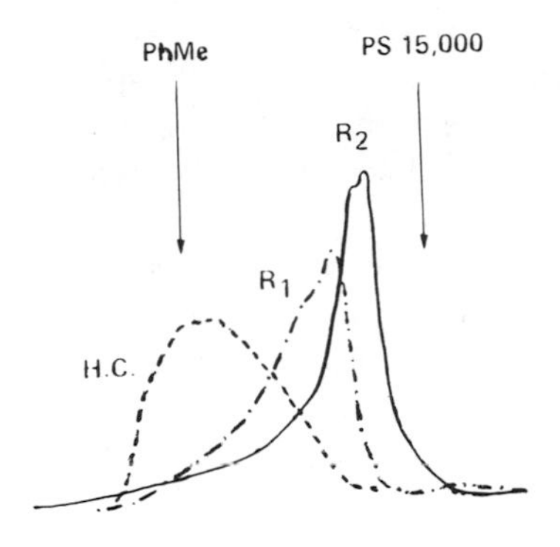

Figure 6. Gel permeation chromatograms of Athabasca maltene fractions separated on Attapulgus clay into hydrocarbons, resins-I (R_1), and resins-II (R_2): columns, 30 cm × ½" μ-Styragel 500 + 30 cm × ½" μ-Styragel 100; solvent, CH_2Cl_2; UV detector; PhMe = V_n of toluene; PS (15,000) = V_0.

on a Styragel 1000-Å column. The response of the refractive index (RI) detector is recorded in both cases. Calibration of the actual distribution was done by repeating the separation ten times and collecting fractions in regular intervals. The individual fractions from the ten consecutive experiments were then combined, the solvent evaporated, and the gravimetric distribution plotted. This experiment has shown two important results. First, the gravimetric measurements tallied well with the elution profile obtained with the RI detector for both the resins and asphaltenes, showing that specific response of RI for these fractions of Athabasca oil sand bitumens is, for practical purposes, sufficiently uniform and proportional to concentration. Second, the gravimetric measurements have shown considerable tailing for both resins and asphaltenes, which was more pronounced for the latter material. These findings clearly demonstrate that for every new type of material, the nonideality or skewness of these curves has to be checked before any attempts at absolute calibration are made. Furthermore, these two curves clearly show a larger than 60% overlap of the resin and asphaltene curves.

Figure 6 represents combined chromatograms of the hydrocarbon fraction and the two resin fractions, R_1 and R_2, as obtained for Athabasca material by conventional SARA procedure. These GPC separations were done on a combination of 500 Å and 100 Å μ-Styragel columns. The scale was roughly calibrated with polystyrene 15,000 (completely excluded) and toluene. A comparison with Figure 5 shows that R_2 would largely tally with the asphaltene curve. This figure also shows that, even at a high column efficiency, the separation by size of R_1 and R_2 is rather poor. Also, the naphthenic hydrocarbons remain virtually unseparated from R_1 while the extension of the hydrocarbon curve beyond the retention time of toluene can only be attributable to polycondensed aromatic hydrocarbon with very short sidechains. Measurements on a number of unsubstituted and substituted polycondensed aromatics in 100-Å μ-Styragel column have shown, in accord with Oelert and Weber (*25*), that their retention times were increasing with the number of aromatic condensed rings while increasing the bulk of the peripheral substituents shifted the retention times towards lower values (*26*).

On the other hand, the latter chromatogram (Figure 6) is the UV record of curves, and we know that at the wavelength used (254 nm) the response dramatically increases with increasing ring condensation, so that the actual mass contributions of all peak tails are apparently much less than indicated by the UV.

Figures 7a and 7b are UV recordings, on the same combined columns (μ-Styragel 500 + 100 Å) and at comparable concentrations, of the various resin fractions isolated in the separation of Cold Lake bitumen resins on an anion (Figure 7a) and a cation (Figure 7b) exchanger column. These curves were compared with MW measurements for these fractions by VPO from methylene chloride (Table VI). The various degrees of sample polydispersity of these fractions are noteworthy. However, these chromatograms have one striking feature that is revealed by the comparison of the base fractions from the cation exchanger with the curve of the nC_5 eluent from the anion exchanger and, further, the relative positions of the acid and base curves with the MWs of these fractions obtained from VPO measurements in the same solvent.

First, the leading edges of all the base peaks agree well with the leading edge of the nC_5 eluate from the anion exchanger. This agrees with our expectation, since the nC_5 eluate from the anion exchanger is a combination of bases and hydrocarbons not retained on the anion exchanger. However, upon comparing the peaks of acids and bases with the MW data (done this time in the same solvent as used for GPC separations), the highest MW by VPO was measured for fraction 15-4 [i.e. benzene/methanol/*i*-propylamine eluate from the cation exchanger (Figure 7b)], presumably strong bases. But, the GPC curve lies well behind that of the benzene/methanol/carbon dioxide eluate from the anion exchanger (Figure 7a), for which a value of 960 was obtained by VPO. As the solvent was the same in both experiments, this discrepancy might have been caused either by the fact that the concentration needed for VPO measurements exceeds that used for GPC (various degrees of association, which will be discussed further below), or that there is a considerable difference in hydrodynamic volumes of bases and acids. No experiments purporting to explain this discrepancy have been carried out. Also, the bases in general tended to higher MWs by VPO than their acidic counterparts (weak, medium, strong).

For the sake of completeness a series of Athabasca asphaltene-derived bases was chromatographed using the same column system and solvent sequence for elution and compared with the bases from the resin fraction of the same bitumen. These results are shown in Figure 8. First, the highest MW fraction of the asphaltenes was the tetrahydrofuran/*i*-propylamine fraction (6500 by VPO). This is one of the fractions that could not be eluted from the column by benzene/methanol/*i*-propylamine. Apparently, tetrahydrofuran/*i*-propylamine is hardly a more polar solution than benzene/methanol/*i*-propylamine. In cation exchanger chromatography, the decisive component

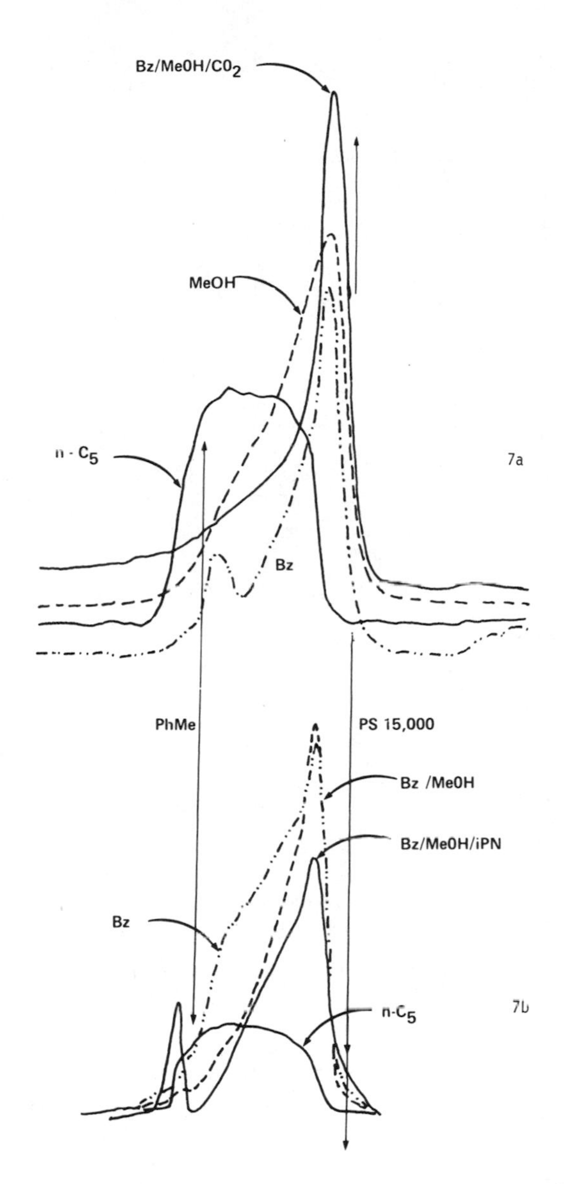

Figure 7. GPC profiles of fractions of Cold Lake resins separated on anion exchange resin (IRA-904) and cation exchange resin (A-15): (a) acid fractions, (b) base fractions; column and conditions same as in Figure 6

should be the amine that interacts with the sulfonic groups of the cation exchanger, thereby displacing the sample. This suggests that there are sorption and/or solubility problems associated with this fraction. Second, comparison of tables of analytical results for the two series of bases in Table VI (MWs) shows the same trend in MW for both series, although data obtained for the

Table VI. Elemental Analyses, Molecular Weights, and Element Ratios for Fractions from Cold Lake Resin Separation on Ion Exchangers

Sample	*C*	*H*	*N*	*O*	*S*	*Molecular Weight*[a]	*Ash*	*H/C*	10^{-3} *N/C*	10^{-2} *O/C*	10^{-2} *S/C*
Maltenes	83.7	10.94	0.3	1.0	4.1	521	0.0	1.57	2.97	0.92	1.81
Acids											
904-1[b]	83.8	11.2	0.2	0.8	4.0	471	0.06	1.61	2.25	0.75	1.79
904-2	81.6	9.0	1.1	2.3	5.6	1050	0.43	1.33	11.66	2.12	2.58
904-3	81.2	9.0	0.9	3.7	5.1	860	0.18	1.33	9.51	3.37	2.34
904-4	79.6	9.8	0.5	5.2	4.4	961	0.39	1.48	5.60	4.92	2.06
Bases											
15-1[c]	84.1	11.1	0.1	1.0	3.8	461	0.00	1.58	1.22	0.86	1.71
15-2	80.8	8.6	1.2	2.7	6.5	887	0.00	1.28	12.84	2.53	3.03
15-3	81.1	9.4	1.3	2.0	6.1	1150	0.00	1.39	13.53	1.87	2.81
15-4	82.3	9.8	1.8	1.3	4.6	1400	0.00	1.43	18.74	1.17	2.11

[a] Solvent CH_2Cl_2.
[b] Anion exchange resin IRA-904 (Rohm & Haas).
[c] Cation exchange resin A-15 (Rohm & Haas).

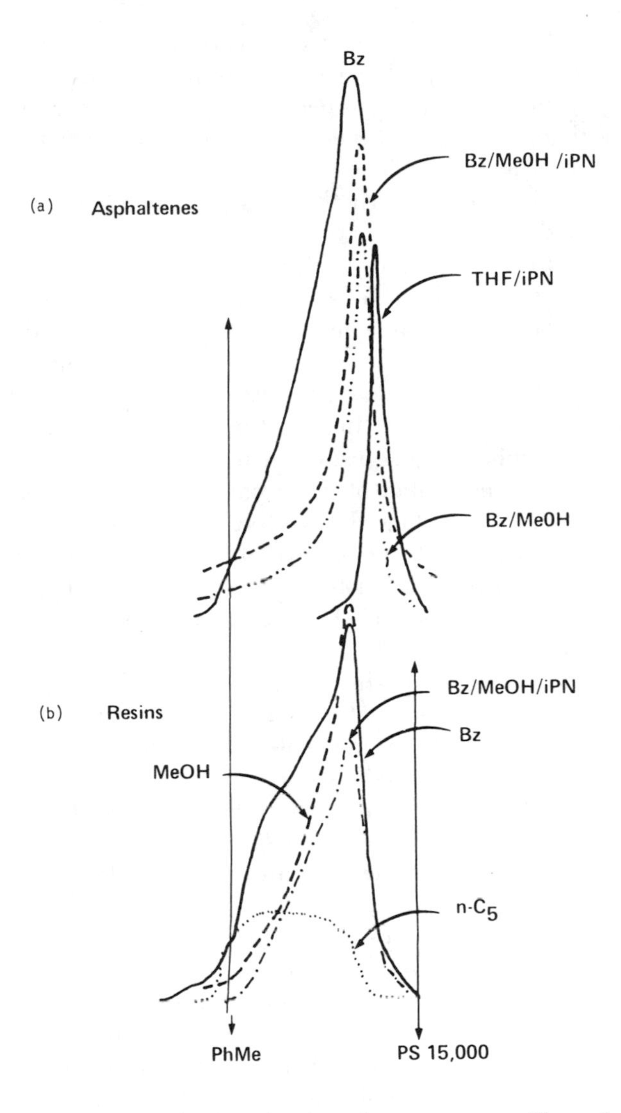

Figure 8. Comparison of gel permeation chromatograms of basic fractions (A-15) of (a) Athabasca asphaltenes and (b) resins: Bz = benzene; MeOH = methanol; iPrN *= iso-propylamine;* nC_5 = n-pentane; *PhMe and PS (15,000) =* V_R *of toluene and* V_0*, respectively*

asphaltenes are about three to four times higher, the GPC curves do greatly overlap. As the exclusion limit has been determined in both cases (as shown in the figures) and as there is very little difference between the GPC curves for the like fractions, various degrees of sorption might explain the closeness of the peaks even in the GPC columns. Only the tetrahydrofuran/*i*-propylamine asphaltene fraction is nearly fully separated from any of the basic resin fractions. The benzene eluate, suggested as a weak basic eluate, is the most polydisperse for both the asphaltene and the resin fractions, suggesting that the selectivity of ion exchanger chromatography under the above described conditions is rather poor for the so-called very weak bases. Essentially the same also holds for the weak acids.

These results show quite clearly that in GPC, which involves a combined size separation and, to a lesser extent, sorption, the difference between resins and asphaltenes is rather marginal in terms of MW, and also that the same holds for fractions separated on the basis of another chromatographic principle, sorption on silica gel and subsequent separation in another dimension by GPC. These results, together with what we have reported concerning asphaltene precipitation using first a poor solvent (nC_5) and then a better solvent (nC_7) to reprecipitate the sample, and also the results obtained by Brulé (*27*) upon fractional precipitation of asphaltenes (such that a limited volume of nC_7 was added to the benzene solution of the bitumen and any precipitate formed removed by centrifugation, followed by adding an incremental volume of nC_7 then another centrifugation, etc.), it seems that for more detailed characterization of the high MW asphaltenes none of the separation methods is substantially better than any other method. The separation of asphaltenes based only on solubility yields the most polydisperse aggregates, and this separation is dependent on the system as a whole, including the soluble portion of the bitumen. On the other hand, gel permeation chromatography of more concentrated sample solutions (approximately 1%) will afford fractions essentially limited by size, and the study of these samples by back-up techniques (NMR, IR, total analysis) should give better information on the development of structural features across the size profile, since by making the GPC fractions sufficiently narrow the extent of polydispersity should be considerably reduced, thus making the averaged values, mostly number averages as determined by VPO, more meaningful.

As we have shown previously in Reference *10*, and as was later confirmed by the work of Brulé (*27*), the more concentrated solutions of the high MW material (1% or more) are, the more polydisperse are the molecular aggregates. We showed that very dilute asphaltene solutions (0.05, 0.01% or less) give gel permeation fingerprints different than those of the more concentrated solutions. Thus, the GPC chromatograms of the various Athabasca asphaltene fractions from ion exchanger separations gave fingerprints shown in Figure 9 for the fractions from the anion exchanger. Only the last two fractions (methylene chloride/carbon dioxide and benzene/acetic acid) appear to be essentially monodisperse while the whole nC_5 asphaltenes and all other fractions are clearly polydisperse, even to such extent that a number of individual peaks can be observed in the chromatogram. This becomes apparent starting with the benzene eluate down to the benzene/carbon dioxide eluate. Again examination by other methods, IR spectrometry in this case, has shown that for the methanol/carbon dioxide fraction a clear selective distribu-

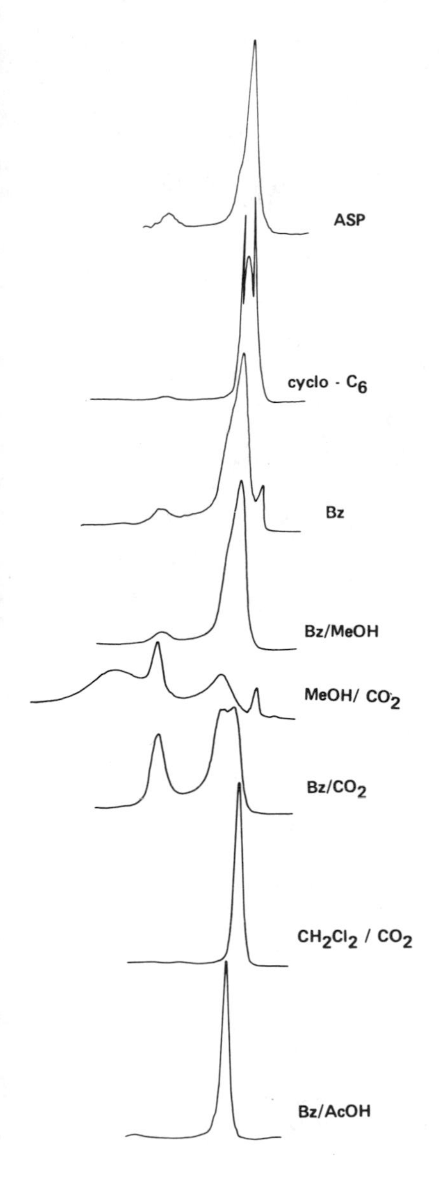

Figure 9. GPC of acidic fractions of Athabasca asphaltenes: 50 μl, 0.05% solution; column, μ-Styragel 1000; solvent, CH_2Cl_2

tion of material exists. To achieve this, the GPC (0.05% Athabasca asphaltene solution) was done four times, and each of the four peaks in the chromatogram shown in Figure 9 was individually collected and concentrated to form a methylene chloride cast for Fourier transform IR spectrometry. From 1000 to 2000 scans sufficed to obtain legible spectra, which are shown in Figure 10. These spectra have been prepared in the absorbance mode so that the largest common peak (centering around 2900 cm^{-1}) was expanded to full scale size. In this way the absorbances in the various functional group regions are

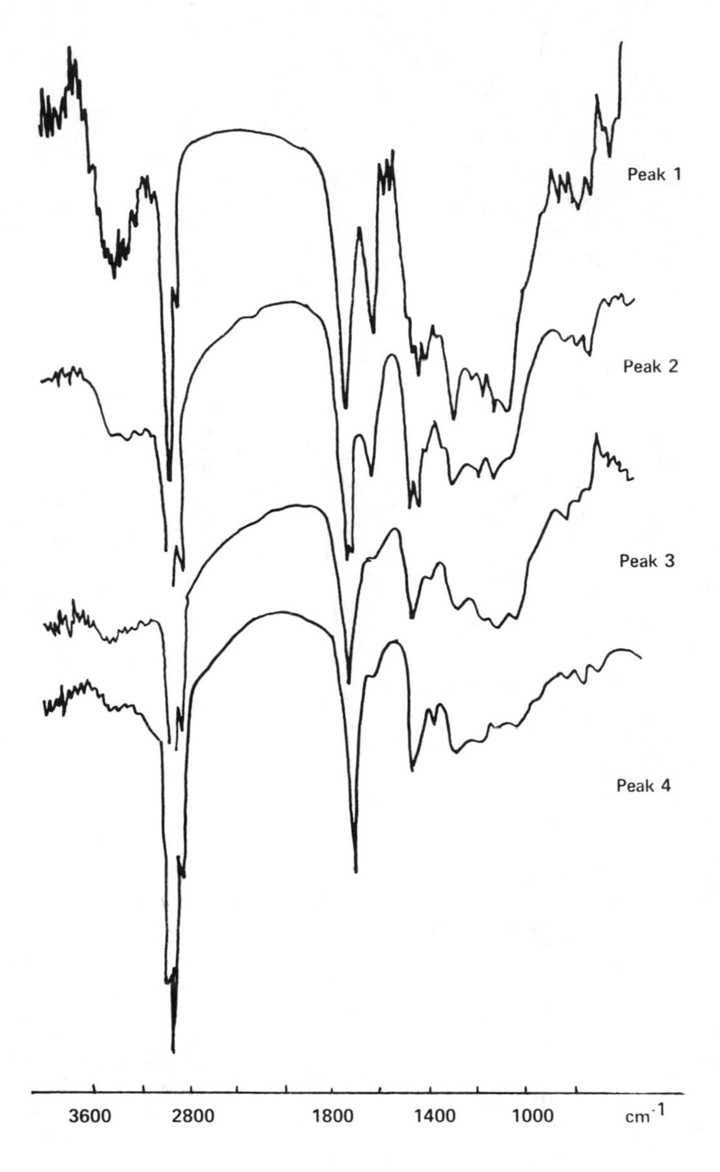

Figure 10. FT–IR spectra of fractions collected from the individual peaks of the MeOH/CO_2 fraction from Figure 9

proportional to the total number of methylene + methyl groups + sample aromaticity, so a rough estimate of functionality with respect to the carbon skeleton can be made. In Figure 10 the peak with the shortest retention time is marked #1 and the other peaks are numbered successively with increasing retention time. The hydrogen bonding region (ca. 3600–2500 cm^{-1}) reveals a large, relatively sharp peak in the first spectrum. This peak decreases over #2 and #3 and is smallest in #4; at the same time, it is becoming progressively broader. The two peaks in the 1800–1600 cm^{-1} region appear with an increasing contribution of the peak at higher wavenumbers to the doublet, until, in #4 the peak at the lower wavenumber has all but disappeared while the other peak has become much sharper. On the other hand, the proportion of methylene groups with respect to methyl groups (1450 cm^{-1} and 1380 cm^{-1}) is changing from about equal proportions in the first two spectra to a ratio, more frequently seen with these types of sample, of 2:1. At the same time, the combined absorption over the 1250–1000 cm^{-1} range is becoming less articulated and smaller, hand in hand with the decreasing hydrogen bonding absorption. In addition, changes in the fingerprinting region are observable. This type of spectra obtained on the various GPC peaks will be treated in detail elsewhere. In this context it suffices to state that: legible spectra can be obtained in FT–IR of these trace amounts of samples with a reasonable number of scans (1000–2000); and preferential concentration of various functions could be observed in the individual GPC peaks. Thus, this method is sensitive enough to gain additional information on the fractions obtained. Molecular weight measurements, elemental analysis, or NMR spectra would require excessive time for sample acquisition owing to their low sensitivity, but in principle these properties can be measured on such fractions.

In another experiment the same Athabasca asphaltene sample was separated on ion exchangers, using a different series of eluents. The fingerprints of the fractions obtained are shown in Figure 11. Here, cyclohexane was omitted from the elution sequence for reasons explained in the section on ion exchanger chromatography. These chromatograms have been recorded at a higher recorder chart speed, using exactly the same system and conditions as before. This time the fingerprint of the benzene fraction was trinodal and the chromatograms of all the fractions were different than the previous series. Only one fraction, the benzene fraction from the cation exchanger (i.e., the most neutral material) showed a shape nearly identical with that of the whole asphaltenes. This difference was surprising and prompted a further study. The only difference (except for the change in the eluent series, which was felt not to be of decisive importance for the overall shape of the curves) was that in the first series the samples were dissolved and immediately run, while in the second series the samples were dissolved on one day and the separation was run on the next day. This suggested some kind of time-dependent sample behavior. Therefore, a more rigorous experiment was set up. The individual samples were freshly dissolved in methylene chloride, filtered, purged with nitrogen, injected onto the GPC column, and after another purging with nitrogen, the sample was sealed in a vial and placed in the drawer to suppress possible light-induced changes or oxidation (even traces of oxygen might be significant at these low asphaltene concentrations). The samples of the whole asphaltenes and of their various fractions from the ion exchangers were freshly dissolved and the first GPC chromatogram was immediately recorded. The individual peaks in the chromatogram were preparatively collected; the separation was repeated three to five times to obtain enough sample. The individual fractions were reinjected on the column and the separation run under identical conditions, as shown in Figure 12. This was done because

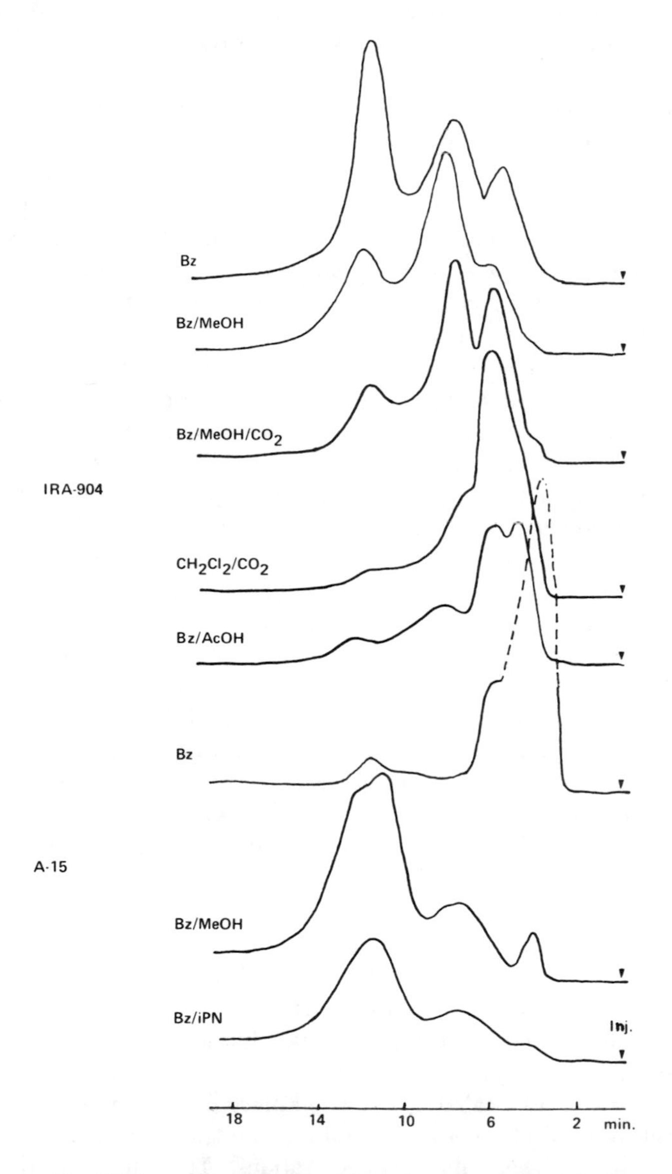

Figure 11. GPC profiles of acid and base fractions from Athabasca asphaltenes (0.05% solution)

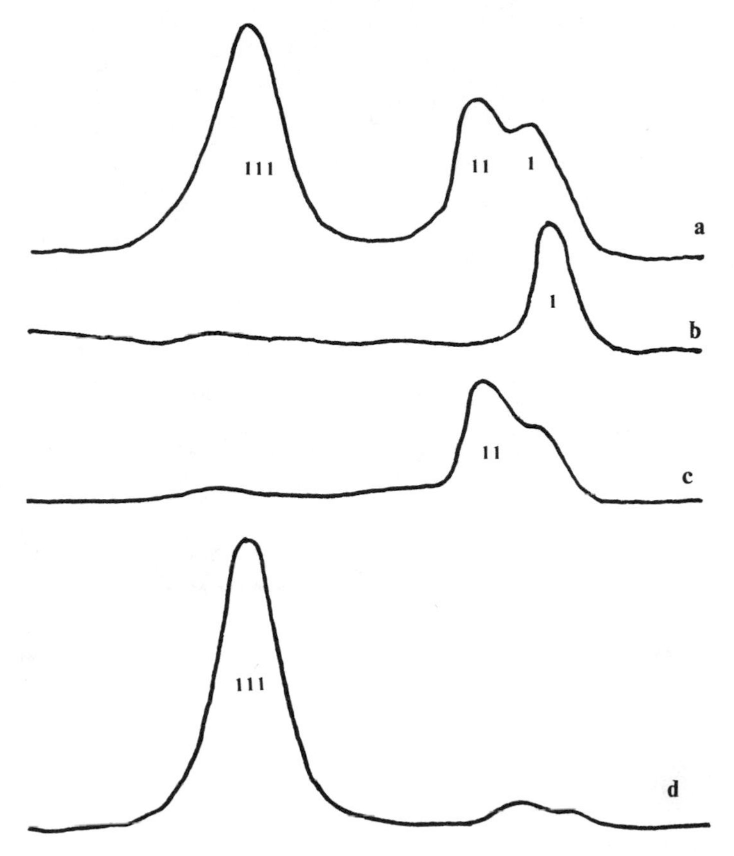

Figure 12. GPC of an Athabasca asphaltene sample (a), and rechromatography of fractions collected as "peaks" I (b), II (c), and III (d)

artifact formation is sometimes experienced (ghost peaks) that can be traced back to a few possible causes. Such frequent checks are important since they confirm the validity of the results. After each injection the sample was flushed with nitrogen, resealed, and stored as described. Repetitive GPC experiments were made for the various samples at intervals shown in Figures 13, 14, and 15. To make the changes in chromatograms more obvious, the area of the largest peak was set equal one hundred (arbitrary units) and the areas of the other peaks were calculated as fractions of one hundred; the individual peaks were plotted in the form of a bar graph at the respective retention times. With these low sample concentrations (0.05–0.01), the UV detector at 254 nm was preferred to the RI detector. Where peaks were not fully separated, a vertical has been dropped at the respective valley, and the areas were not further corrected. The integration was done by counting the number of squares on the chart paper since common integrators cannot be employed for this type of chromatogram.

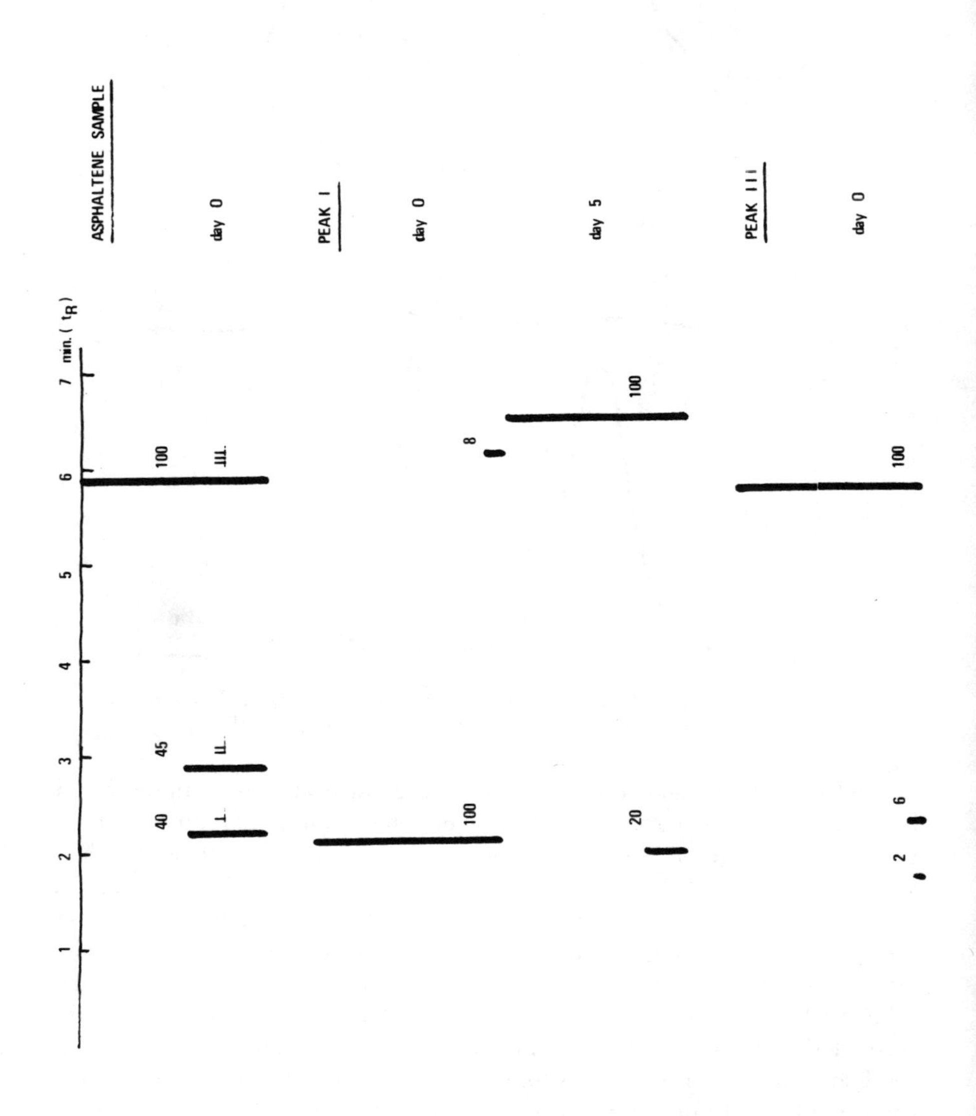

ASPHALTENE SAMPLE
day 0
PEAK I
day 0
day 5
PEAK III
day 0
min. (t_R)
1
2
3
4
5
6
7
100
III
45
II
40
I
8
100
100
20
100
6
2

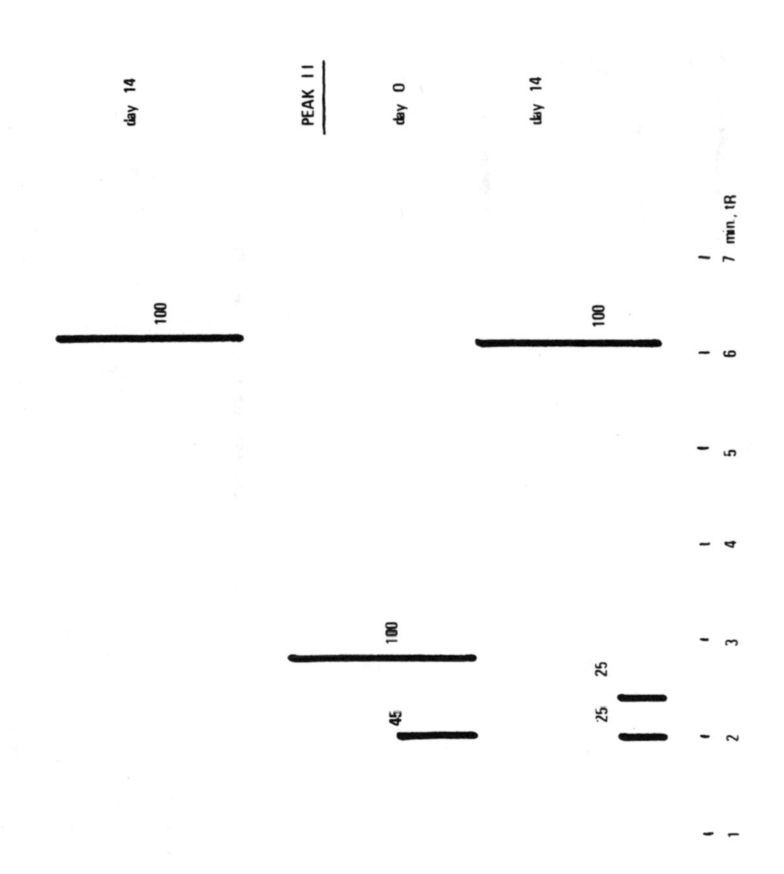

Figure 13. Time study of changes in GPC profiles of fractions from Figure 12

The first series of experiments (Figure 13) refers to the individual peaks of total asphaltene separation shown in Figure 12. Thus, Peak I was nearly completely converted into a low MW species in only five days; Peak III shows only a slight decrease in molecular size. Peak III is the lowest MW peak in the chromatogram (Figure 12). On the other hand, Peak II largely shifted to low MW, while some redistribution occurred in the high MW region.

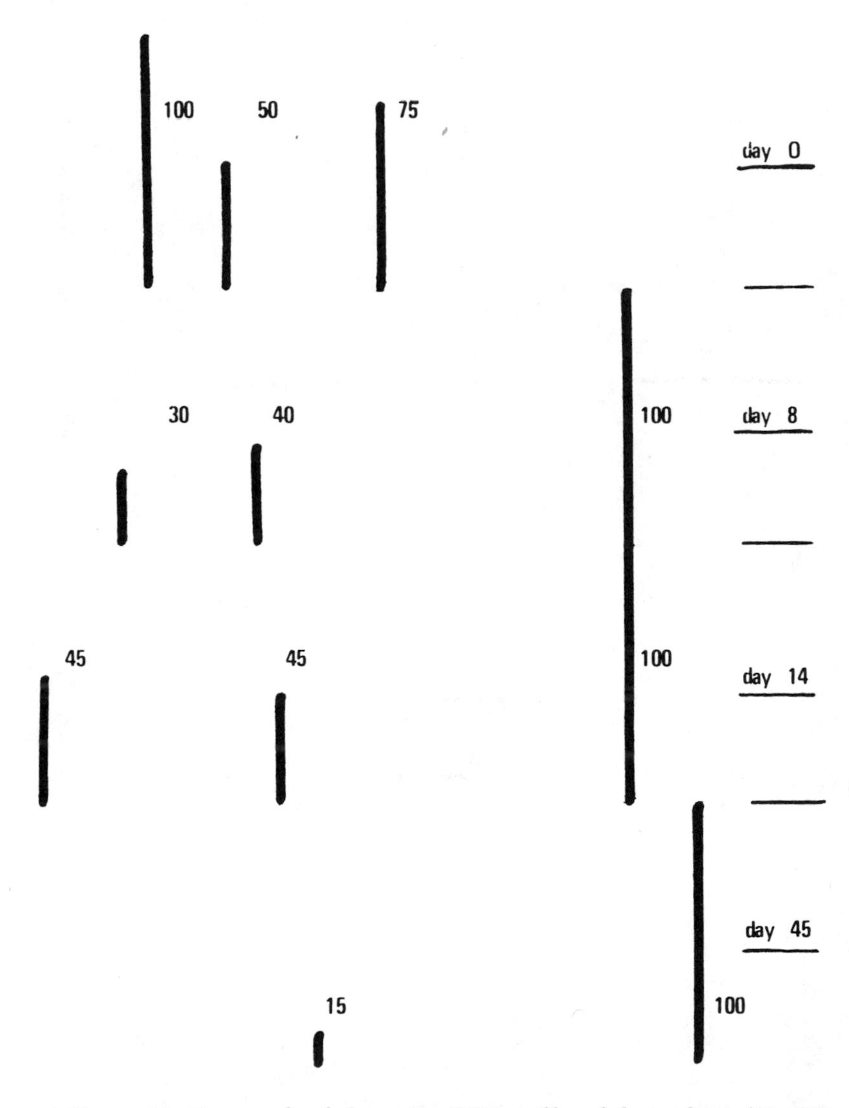

Figure 14. Time study of changes in GPC profiles of the acidic Bz/MeOH fraction of Athabasca asphaltenes

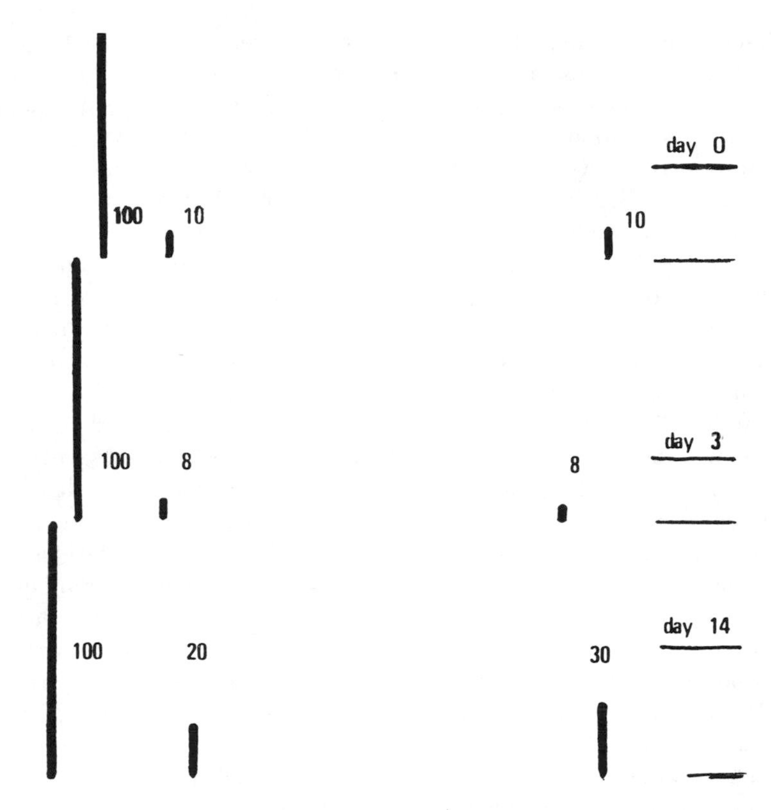

Figure 15. Time study of changes in GPC profiles of the Bz fraction of Athabasca asphaltene

The peaks of the chromatogram in Figure 14 of the benzene/methanol fraction (Figure 12, Peak II) from the anion exchanger redistributed as shown. This series of chromatograms shows the slowness of the changes; complete conversion to low MW species took forty-five days. This series of chromatograms of the acidic fraction can be compared with the changes in the shape of the GPC chromatogram of the benzene eluate from the anion exchanger (*see* Figure 9). Unlike the acid fraction, the benzene fraction changed very little over a period of fourteen days, while with slightly increasing size of Peaks II and III the high MW peak was shifted to slightly shorter retention times.

These few experiments did not permit us to gain a deep understanding of the phenomena involved. However, they did illustrate the complexity of the asphaltene aggregate and the instability of this complex system. Asphaltenes and their fractions tend to give molecular species of much lower MW on dilution, and this process is time dependent. This process of unravelling the micelle can be very slow. The sample solutions were examined microscopically for traces of precipitate, but no such traces have been discovered. Therefore, it seems that the whole samples undergo these shifts to considerably lower MW species, and the slowness of the processes suggests some sort of kinetically controlled dissociation.

Since classical *R*—*S*—*R* or C—C bonds do not dissociate (and would not dissociate even if, e.g., oxidation were taking place), it seems to us that all formulations of asphaltenes as "poly-polymers" [e.g., Yen (*7*), or —S— polymers, e.g., Ignasiak et al. (*8*)] give an impression of greater stability than is exhibited by the decomposition of the aggregates. More detailed studies of these time-dependent phenomena could shed more light on the mechanisms of aggregation and disaggregation of these species of bitumens characterized by their high apparent MW.

Brulé (*27*) deduced from his experiments on asphalts that "isolated asphaltenes form agglomerates which may be dissociated into micelles by simple dilution and that the intensity of the phenomena should characterize the force of interaction." He also noticed that in some cases "there is a drift towards small sizes which implies a continual departure by molecules of all dimensions." On different types of material there was a general trend towards molecules in the approximate range of 50 Å; moreover, our experiments have clearly demonstrated that some of the processes involved may be very slow. Also, the redistribution of material across the chromatogram might be attributable to various types of reassociation once some part of the aggregate dissociated to small MW species.

These results suggest that even if the model of Dickie and Yen (*21*) might explain some of the phenomena, the asphaltenes are a system with dynamic behavior depending on all system components; these experiments and the precipitation experiments of Boduszynski (*6*) are suggesting the same thing, that we do not yet understand the actual nature of asphaltenes.

In the samples from preparative separations of more concentrated solutions of asphaltenes and resins, the only truly continuous change is recorded for the number of spins/gram of material (*28*). It is quite possible that the stable radical nature of these materials—asphaltene values are in the range of 10^{-17}–10^{-18} spins/g, resins 10^{-15}–10^{-17} spins/g (evacuated)—is the main contributor to such associations, while dissociation occurs through separation of charge-transfer complexes into progressively simpler components. Classical dissociation of the ionic type or separation of hydrogen-bonded species upon dilution should be only a minor contributor because (from elemental analyses) the number of functional groups per molecule is essentially very small and also because GPC experiments with dilute solutions and isolation of high MW fractions from sorbents appear to be the least polar material. With increasing polarity (IR evidence) the MW became progressively lower. This combined evidence demonstrates that the least polar material sometimes forms the largest agglomerates and that these agglomerates are relatively the most stable.

Conclusions

The separations of resins and asphaltenes under comparable conditions done on the fractions from a common source—Athabasca (in some cases Cold Lake) bitumen—have confirmed a number of postulates that have appeared in the recent literature.

First, in all three systems studied (adsorption, ion exchanger, and gel permeation chromatography) there is a considerable overlap in behavior of the resins and asphaltenes. As the separation principle is essentially different in all three cases, these similarities clearly show that resins and asphaltenes are part of a continuum of structures and confirm that solubility separations give complex aggregates where composition may vary with the precipitation conditions. The adsorption separations show that it is not necessarily the presence of functional groups that results in high MW; the ion exchanger separations add to the picture by showing that the distribution of acidic and basic components in the two fractions is similar. Gel permeation chromatography of more concentrated solutions (1% or higher) demonstrates a continuum of increasing MWs.

Second, changes in solubility of fractions separated from these materials suggest that upon solvent evaporation new types of agglomerates are forming after removal of some mixture components. This implies that the agglomerating or associating entities are much smaller than thought previously.

Third, GPC of very dilute solutions of asphaltenes and their fractions clearly confirm the above implications and demonstrate that these materials represent a dynamic, time-dependent system that changes to MW entities of the size of polyaromatic hydrocarbons and similar materials. This in turn indicates that the forces involved in the apparent high MW species formation are not fully understood and questions the tenability of formulations based on σ-bonded polymers to explain the apparent high MW of the aggregates.

Even though precipitation of asphaltenes can be source of valuable information to the processing chemist, it is unfortunate from the viewpoint of the analytical chemist. The precipitation furnishes polydisperse mixtures whose degree of polydispersity is not known because most of the methods employed for the instrumental backup and for MW measurement are of the number-averaging type. Since size and polarity both contribute to the degree of precipitation, methods based on separation by size, followed by separations based on polarity, should bring some simplification into the fractionation system. On the other hand, separation methods are also limited in their scope, which, combined with incipient solubility problems (i.e., limitations on the variety of available solvents of increasing polarity), makes the study of the high MW fractions of bitumens a difficult task.

Acknowledgment

The financial support of this investigation by the Alberta Oil Sands Technology and Research Authority and the Canada Alberta Energy

Resources Research Fund is gratefully acknowledged. The authors also wish to express their appreciation to Dr. D.S. Montgomery for helpful suggestions in the preparation of the manuscript.

Literature Cited

1. Bunger, J. W.: Cogswell, D. E.; Zilm, K. W. *Am. Chem. Soc., Div. Pet. Chem., Prepr.* (Washington, D.C., Sept., 1979) *24*(4), 1017.
2. Bunger, J. W. *Am. Chem. Soc., Div. Pet. Chem., Prepr.* (New Orleans, Mar., 1977) *22*(2), 716.
3. McKay, J. F.; Amend, P. J.; Cogswell, T. E.; Harnsberg, P. M.; Erickson, R. B.; Latham, D. R. In "Analytical Chemistry of Liquid Fuel Sources," *Adv. Chem. Ser.* ***1978,*** *170,* 128–142.
4. Mitchell, D. L.; Speight, J. G. *Fuel* **1973,** *52,* 149.
5. Koots, J. A.; Speight, J. G. *Fuel* **1975,** *54,* 179.
6. Boduszynski, M. M. *Am. Chem. Soc., Div. Pet. Chem., Prepr.* (Washington, D.C., Sept., 1979) *24*(4), 935.
7. Yen, T. F. *Am. Chem. Soc., Div. Pet. Chem., Prepr.* (Washington, D.C., Sept., 1979) *24*(4), 901.
8. Ignasiak, T.: Kemp-Jones, A. V.; Strausz, O. P. *J. Org. Chem.* **1977,** *42,* 312.
9. Long, R. B. *Am. Chem. Soc., Div. Pet. Chem., Prepr.* (Washington, D.C., Sept., 1979) *24*(4) 891.
10. Selucky, M. L.: Ruo, T. C. S.; Skinner, F.; Kim, S. S. Presented at *Confab*, Saratoga, Wyoming, *July 26, 1978.*
11. Selucky, M. L.; Chu, Y.; Ruo, T. C. S.; Strausz, O. P. *Fuel* **1977,** *56,* 369.
12. Selucky, M. L.; Chu, Y.; Ruo, T. C. S.; Strausz, O. P. *Fuel* **1978,** *57,* 9.
13. Bulmer, J. T.; Starr, J., Eds. "SYNCRUDE, Analytical Methods for Oil Sand and Bitumen Processing"; Syncrude Canada Ltd.: 1979; pp. 121–123.
14. Selucky, M. L.; Ruo, T. C. S.; Chu, Y.; Strausz, O. P. In "Analytical Chemistry of Liquid Fuel Sources," *Adv. Chem. Ser.* **1978,** *170,* 117–127.
15. Selucky, M. L.; Ruo, T. C. S.; Chu, Y.; Strausz, O. P. *Am. Chem. Soc., Div. Pet. Chem., Prepr.* (New Orleans, Mar., 1977) *22*(2), 695.
16. Haines, W. E.; Thompson, C. J. "Separating and Characterizing High-Boiling Petroleum Distillates," The USBM-API Procedure, LERC/RI-75/5 and BERC/RI-75/2, July, 1957.
17. Selucky, M. L.; Montgomery, D. S.; Strausz, O. P., unpublished data.
18. Bunger, J. W.; Thomas, K. P.; Dorrence, S. M. *Fuel* **1979,** *58,* 183.
19. Snyder, L. R. "Principles of Adsorption Chromatography"; Marcel Dekker: New York, 1968; p. 147.
20. Selucky, M. L., unpublished data.
21. Dickie, J. P.; Yen, T. F. *Anal. Chem.* **1967,** *39,* 1847.
22. Reerink, H.; Lizjenga, J. *Anal. Chem.* **1975,** *47,* 2160.
23. Snyder, L. R. *Anal. Chem.* **1969,** *41,* 1223.
24. Altgelt, K. H. *Bitumen, Terre, Asphalte, Peche* **1970,** *21,* 475.
25. Oelert, H. H.; Weber, J. H. *Erdoel Kohle* **1970,** *23,* 484.
26. Skinner, F.; Selucky, M. L., unpublished data.
27. Brulé, B. *J. Liq. Chromatogr.* **1979,** *2*(2), 165.
28. Schulz, K. F.; Selucky, M. L. *Fuel* **1981,** in press.

RECEIVED June 23, 1980.

7

Asphaltenes in Petroleum Asphalts

Composition and Formation

MIECZYSLAW M. BODUSZYNSKI[1]

Department of Petroleum and Coal Chemistry, Polish Academy of Sciences, 1-go Maja 62, 44-100 Gliwice, Poland

The effect of intermolecular associations of polar compounds upon formation of asphaltenes in straight-reduced and air-blown petroleum asphalts is discussed. Two separation schemes are used to evaluate asphalt composition: a separation method based on precipitation of asphaltenes followed by maltene separation on alumina; and a separation method based on removal of acids and bases using anion and cation exchange resins, respectively, followed by separation of neutrals on alumina. Asphaltenes appear to be agglomerates of highly polar compounds formed by intermolecular associations. Field ionization mass spectrometry (FIMS) molecular weight profiles of asphaltenes were found to be similar to those of whole asphalts, from 300 to 1800 molecular weight, and averaging about 1000. Asphaltenes' compositions vary because of the commonly used, chemically nonspecific precipitation method, which is based only upon solubility parameters. The increase in asphaltenes content upon air-blowing of asphalt results from chemical reactions leading to the formation of more strongly associating but not higher molecular weight compounds.

Petroleum asphaltenes are commonly viewed as an undesired component of crude oil that creates serious difficulties in upgrading of petroleum heavy ends. However, it is not often recognized that processing of petroleum residua also includes production of asphalts where asphaltenes are not only a very desired component but the component that determines, to a great extent, the physical properties of an asphalt.

[1]Current address: Chemistry Department, University of Wyoming, Laramie, Wyoming 82071

0065-2393/81/0195-0119$05.00/0

Petroleum asphalts can be manufactured from various crudes using different processes. Asphalt, being the highest boiling fraction of petroleum, can be manufactured by vacuum distillation of crude oil or by treatment of petroleum residue with propane under controlled conditions. Asphalt can also be a product of an air-blowing process in which petroleum residue is contacted with air at temperatures ranging from 200° to 280° C.

The generally accepted primary subdivision of petroleum asphalts, regardless of the process used for their manufacture, is a separation into asphaltenes and maltenes by precipitation with low-boiling saturated hydrocarbons (*1*). Recently the *n*-heptane precipitate has been defined as asphaltenes (*2*). The asphaltene separation is based on the solubility of asphalt components and has limited analytical value because it does not produce chemically defined fractions. Previous work (*3*, *4*, *5*) with maltenes and asphaltenes separated from Romashkino petroleum asphalt showed that the same classes of polar compounds, acids, and bases, were present in both fractions of the asphalt. More recent work (*6*) on high vacuum residue from the same petroleum showed that the compound types that precipitate as asphaltenes are present in the acid and base fractions and that these compound types are capable of association through hydrogen bonding. Intermolecular associations also have dramatic effects on apparent average molecular weight measurements by vapor pressure osmometry. Association results in erroneously high molecular weights of polar components, particularly of asphaltenes.

The purpose of this study is to investigate the composition and the formation of asphaltenes in straight-reduced and air-blown petroleum asphalts. The following topics are discussed: asphalt solubility in normal alkanes; effect of intermolecular associations on average molecular weight measurements by vapor pressure osmometry; and formation of asphaltenes in straight-reduced and air-blown asphalts. The chemical separation method based on reactivity was used to separate asphalts into chemically meaningful fractions consisting of compound classes of different chemical structure and reactivity. The compositional changes in asphalts upon air-blowing, determined by the chemical separation method, were compared with those determined by the precipitation method. The emphasis of this study is on the role of polar compounds capable of intermolecular associations in the formation of asphaltenes.

Experimental

Samples. Two straight-reduced asphalts from Romashkino-type (29° API gravity) crude petroleum, 100 penetration (SR-100) and 125 penetration (SR-125), and three air-blown asphalts of different degrees of oxidation, 100, 50, and 25 penetration (AB-100, AB-50, and AB-25) were investigated. The air-blown asphalts were derived from SR-125. The properties of the asphalts are given in Table I.

Table I. Properties of Asphalts

Properties	SR-100	SR-125	AB-100	AB-50	AB-25
	Asphalt				
Penetration at 25°C 100 g/5 s, 0.1 mm	102	125	97	52	25
Softening point, R&B (°C)	44	37	46	58	76
	Weight Percent in Asphalt				
Carbon	85.3	85.3	85.7	86.0	86.1
Hydrogen	10.6	10.2	10.3	10.4	10.1
Sulfur	3.3	3.7	3.3	2.9	3.0
Nitrogen	0.6	0.7	0.5	0.5	0.4
Oxygen[a]	0.2	0.1	0.2	0.2	0.4
Benzene insoluble (wt %)	<0.1	<0.1	<0.1	<0.1	<0.1
Asphaltenes (C_7) (wt %)	10.6	7	9	17	23
	Weight Percent in Asphaltenes				
Carbon	84.8	85.6	85.7	85.7	85.3
Hydrogen	8.0	7.7	7.3	7.4	7.6
Sulfur	4.1	4.2	4.2	3.9	3.9
Nitrogen	1.2	1.2	1.0	1.0	1.0
Oxygen[a]	1.9	1.3	1.8	2.0	2.2

[a]By difference.

Reagents. Normal pentane (99%), heptane (99%), and decane (99%) were further purified by distillation and by percolation through activated silica gel. Reagent-grade benzene, methanol, and trichloroethylene (TCE) were redistilled. Isopropylamine (IPA) was used as received.

The anion exchange resin was Amberlite IRA-904, and the cation exchange resin was Amberlyst 15. Preparation of the resins was similar to that described by Jewell et al. (7).

Alumina (neutral, Brockmann II) was activated by heating at 400° C for 48 h and was stored in a desiccator.

Precipitation of Asphaltenes. Three different normal alkanes—pentane, heptane, and decane—were used for the asphaltene precipitation. An asphalt sample (SR-100) was agitated with the solvent at room temperature for approximately 30 min and then left for 24 h in the dark. The ratio of 25 mL of the solvent to 1 g of asphalt was used. The solution was decanted from the precipitate, and the asphaltenes were filtered and extracted with the solvent (pentane or heptane), using a Soxhlet extractor, or thoroughly washed with hot (80° C) decane. The asphaltenes were then dried in a vacuum oven under nitrogen and weighed. The pentane asphaltenes were then treated with heptane, and the heptane asphaltenes were treated with decane, using the same procedure as for the precipitation of asphaltenes from the original asphalt. The separation scheme illustrating precipitation of asphaltenes using different normal alkanes is shown in Figure 1.

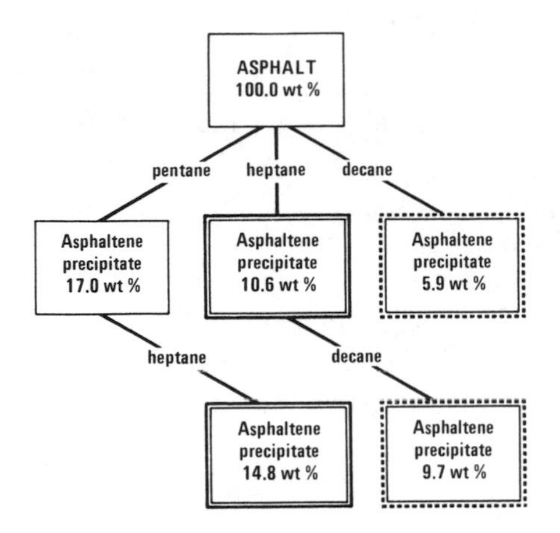

Figure 1. Precipitation of asphaltenes using different n-*alkanes (25)*

Separation of an Asphalt—Method 1. An asphalt was separated into asphaltenes and maltenes using heptane precipitation followed by a separation of maltenes on alumina into a saturate fraction and three aromatic fractions using a procedure (*8*) similar to that developed by Corbett (*9*). The separation scheme is shown in Figure 2.

Separation of an Asphalt—Method 2. Asphalt samples were separated into six fractions: acids, bases, saturates, aromatics-1, aromatics-2, and aromatics-3. Acids were isolated using anion exchange resin, and bases were isolated using cation exchange resin. The remaining fraction (asphalt, less acids and bases) was separated on alumina using the same procedure as in Method 1 for the separation of maltenes. The separation scheme is shown in Figure 3.

CHROMATOGRAPHY ON ANION EXCHANGE RESIN. The sample of asphalt dissolved in benzene was charged into the anion exchange resin column prewet with benzene. A water-jacketed glass column of 1.4 × 120 cm, equipped with a solvent-recycling arrangement that permitted the continuous elution of the sample, was used. The ratio of the sample to the resin was 1:5 by weight. The temperature of the column was maintained at 30° C. Unreactive material was washed from the resin with benzene for approximately 24 h. The reactive compounds (acids) were recovered as two subfractions, the first eluted with methanol for 6 h, the second with trichloroethylene for 6 h. The two acid subfractions were then combined for further study.

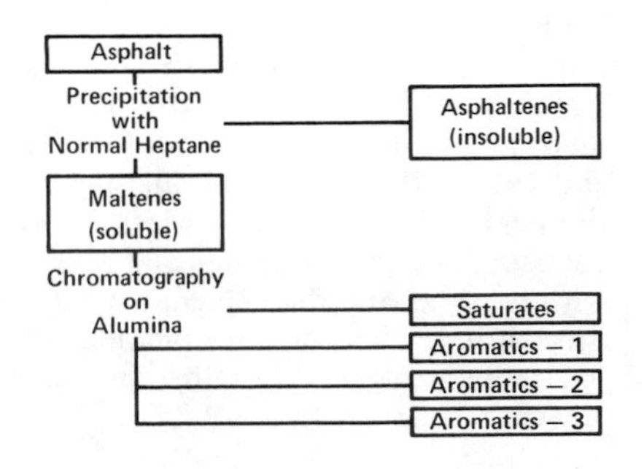

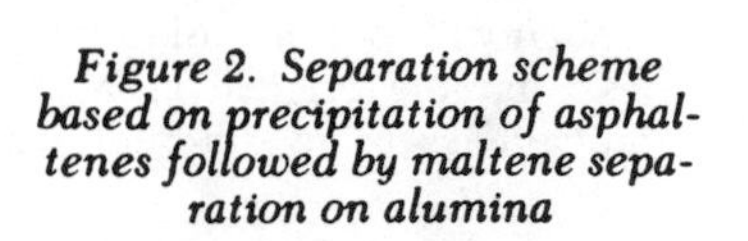

Figure 2. Separation scheme based on precipitation of asphaltenes followed by maltene separation on alumina

CHROMATOGRAPHY ON CATION EXCHANGE RESIN. The sample of acid-free asphalt in benzene solution was charged into the cation exchange resin column prewet with benzene. The size of the column, the temperature, and the ratio of the sample to the resin were the same as in the anion exchange chromatography step. Unreactive material was washed from the resin with benzene for approximately 24 h. The reactive compounds (bases) were recovered as two subfractions, the first eluted with methanol–8 percent isopropylamine for 6 h and the second with trichloroethylene for 6 h. The two base subfractions were then combined for further study.

Gel Permeation Chromatography (GPC). GPC was used to determine molecular-size distributions of asphalts and their fractions. Chromatograms were obtained by means of a Waters Associates ALC/GPC 301 chromatograph equipped with 8.5 × 10^3-, 10^3-, 500-, and 70-Å Styragel columns. The instrument was operated at ambient temperature. Tetrahydrofuran (THF) was used as the solvent. Concentration of the sample in tetrahydrofuran was less than 0.5 wt %. The flow rate was 1 mL/min. Monodispersed polystyrene standards provided by Waters Associates were used to calibrate the system. Thus, weight-average molecular weights (M_w) determined represent relative molecular size only.

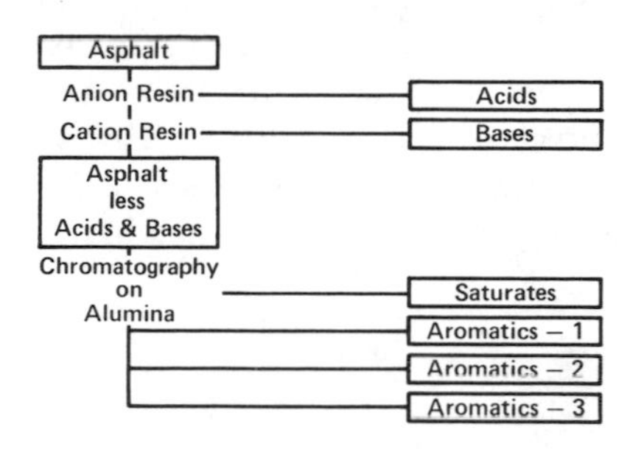

Figure 3. Separation scheme based on removal of acids and bases followed by separation of neutrals on alumina

Field-Ionization Mass Spectrometry (FIMS). FIMS was used to determine molecular weight distributions of asphalts and derived fractions. FIMS spectra were obtained at SRI International using a procedure previously described (*10*). FIMS technique produces molecular (parent) ions that undergo little or no fragmentation. The sample to be examined is volatilized at high vacuum (10^{-6} – 10^{-7} mm Hg) and at about 350° C temperature and is passed through a strong electric field obtained with a needle point or razor blade edge. Electrons escape the molecules, producing positive ions, but no chemical bond cleavage occurs. The spectrum obtained by repeated multiscanning represents a molecular weight profile of the original sample.

Asphalt samples (SR-125 and AB-25) were 100% volatile under conditions of analysis, and asphaltene samples (asphaltenes SR-125 and asphaltenes AB-25) were 92% and 83% volatile, respectively. No cracking, pyrolysis or other form of molecular decomposition was observed.

Vapor Pressure Osmometry (VPO). Apparent average molecular weights were determined using VPO in benzene solution at concentrations of 2–20 g/L. Results obtained at three different concentrations were then extrapolated to infinite dilutions to determine the average molecular weight.

Elemental Analyses. Carbon, hydrogen, and sulfur were determined gravimetrically as carbon dioxide, water, and barium sulfate, respectively. Nitrogen was determined by a micro-Dumas method. Oxygen was determined by difference.

Results and Discussion

Asphalt Solubility in Normal Alkanes. The separation of an asphalt into two fractions—asphaltenes and maltenes—by precipitation with low molecular weight alkanes is a physical method of separation based on solubility. Figure 1 shows that pentane precipitates more asphalt components (17.0 wt %) than does heptane (10.6 wt %). It would be expected that when pentane asphaltenes are treated with heptane, the amount of material equal to 10.6 wt % (based on asphalt) would be precipitated. However, more—14.8 wt %—is precipitated. A similar, but even more pronounced, effect can be seen when heptane and decane treatments are compared (*see* Figure 1).

The data in Figure 1 indicate that the amount of asphaltene precipitate is not only dependent upon the kind of an alkane solvent but is also affected by the initial composition of the material from which it is precipitated. Treatment with normal alkane upsets the solubility equilibrium of a very complex mixture of various compounds comprising asphalt or its fraction. As suggested by McKay et al. (*11*) and also by Long (*2*), solubility equilibrium is maintained by two major factors: polarity and molecular weight. According to these authors, less polar materials of higher molecular weight and more polar materials of lower molecular weight both precipitate as asphaltenes. However, it has been shown (*6*) that all compound types present in the asphalt have about the same molecular weight distribution. Thus, it seems that solubility of

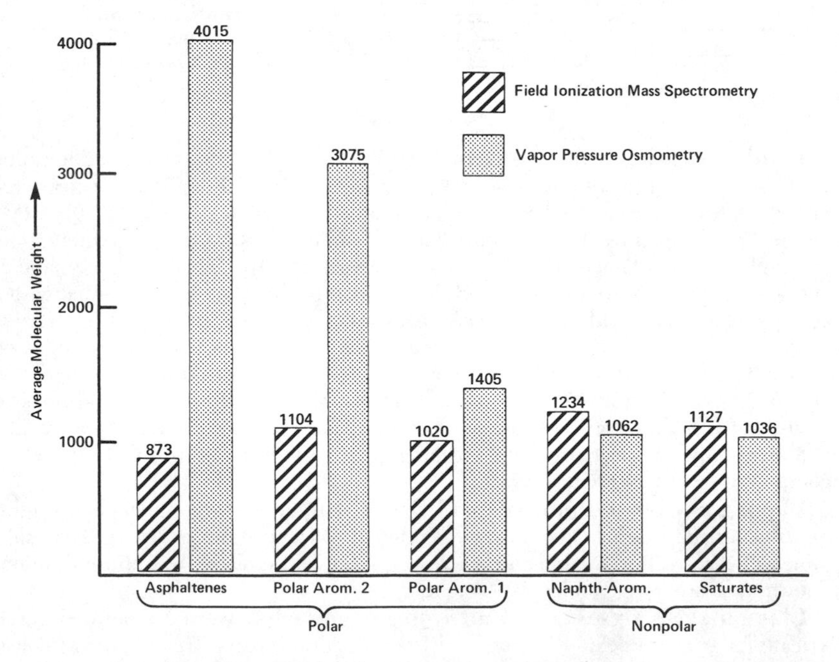

Figure 4. Effect of intermolecular association on apparent average molecular weight

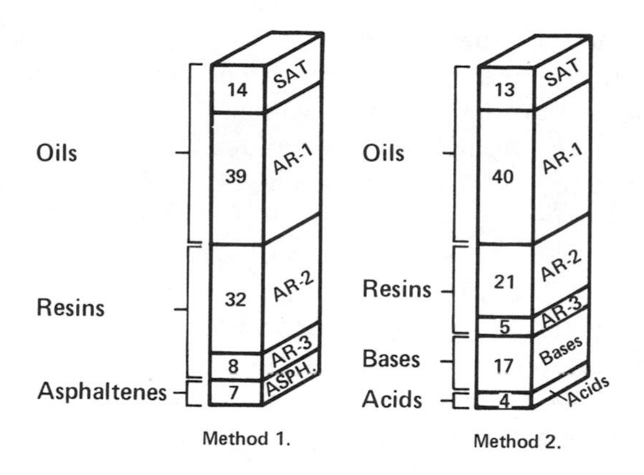

Figure 5. Composition of straight-reduced asphalt (SR-125)

asphalt components in low-boiling alkanes is dependent upon molecular interactions of polar compounds, resulting in the formation of aggregates of limited solubility.

Effect of Intermolecular Associations on Average Molecular Weight by VPO. The problem of determining the molecular weight of asphaltenes was discussed by Speight and Moschopedis (*12*). It has been reported (*13*) that average molecular weights of asphaltenes determined by VPO are dependent not only on the kind of solvent but also on the solution temperature. Recent study (*6*) has shown that intermolecular associations lead to erroneously high VPO molecular weight values, not only for asphaltenes but also other polar fractions of the asphalt. Figure 4 illustrates the effect of intermolecular associations on VPO average molecular weight of generic fractions separated from the straight-reduced asphalt using Corbett's procedure (*9*). Apparent average molecular weights obtained by VPO are compared with those determined by FIMS. A dramatic effect of intermolecular associations on VPO average molecular weight of asphaltenes is evidenced by comparison of VPO and FIMS data (4015 and 873, respectively). This effect diminishes with the decrease in polarity of the fractions. Polar aromatic fractions, particularly polar aromatics-2, show considerable differences in average molecular weights determined by the two methods, while nonpolar saturates and naphthene-aromatics show practically no differences.

Formation of Asphaltenes in Straight-Reduced Asphalt. The straight-reduced 125–penetration asphalt (SR-125) has been separated into fractions using two separation procedures illustrated in Figures 2 and 3. The results of the separations are compared in Figure 5. The comparison of the amounts of fractions generated from the asphaltene-free (maltene) portion of SR-125 asphalt (Method 1) and acid- and base-free portion of SR-125 asphalt (Method 2) by chromatography on alumina shows that the amounts of saturates plus aromatics-1, often called oils, remain constant (53 wt %). Differences can be observed in the amounts of polar fractions (resins) defined as aromatics-2 and aromatics-3. The prior separation of acids and bases directly from the asphalt

(Method 2) results in the reduction of the amount of aromatics-2 and aromatics-3 (resins). The acid and base fractions also appear to contain asphalt components defined by Method 1 as asphaltenes.

Recent work (*6*) on high vacuum residue from the same petroleum as used in this study showed that the compound types comprising acid and base fractions include pyrroles, phenols, amides (including 2-quinolones), carboxylic acids, and pyridines. These are compounds capable of association through hydrogen bonding (*14, 15, 16*). Barbour and Petersen (*17*) reported that "considerable hydrogen bonding occurs within asphalt between naturally occurring hydrogen-bonding acids and bases."

The results of this study imply that the asphaltenes present in the straight-reduced asphalt primarily consist of compounds capable of hydrogen bonding, that is, acids and bases.

Formation of Asphaltenes upon Air-Blowing of Asphalt. It seemed worthwhile to investigate the mechanism of asphaltene formation when a straight-reduced asphalt is subjected to the air-blowing process. The straight-reduced 125-penetration asphalt was air-blown to produce three penetration grades of asphalts: AB-100, AB-50, and AB-25. Asphaltenes are formed during the air-blowing process. Figure 6 illustrates increase of asphaltene content in asphalts studied. The mechanism of asphaltene formation is not known, although many attempts have been made to understand the mechanism (*18*). During air-blowing of an asphalt, some oil components are converted into resins and then into asphaltenes. However, dehydrogenation and condensation reactions are involved and oxygen is not added to the asphalt product except in minor amounts (*19, 20, 21*).

Molecular Weight Distribution of Asphalts Upon Air Blowing. In this study, investigations into the effects of air-blowing upon the molecular weight of asphalts and their fractions were conducted using GPC and FIMS.

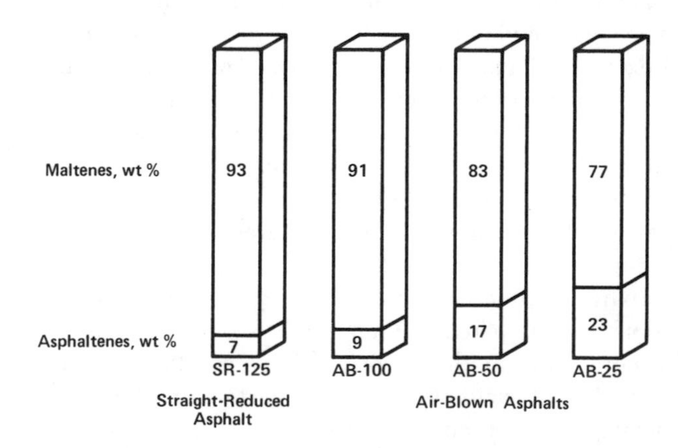

Figure 6. Asphaltene formation upon air-blowing of asphalt

GPC was used as a major tool to determine molecular size distributions of asphalts and their fractions. However, one must bear in mind that data obtained by GPC do not represent actual, absolute, molecular weight values of asphalt compounds since the system has been calibrated using polystyrene standards. Besides, it is known that factors such as the adsorption of polar compounds on the gel and/or intermolecular associations of polar compounds can affect GPC results. Each factor affects the results in a different way. Adsorption on the gel would result in lower apparent molecular size values, while the association of these compounds would cause earlier elution, giving apparent molecular size values higher than the actual values. The effect of intermolecular associations in asphalts on GPC data, even at concentrations usually considered low enough to destroy all solute–solute associations, has been reported in the literature (*22, 23*).

GPC chromatograms of the straight-reduced SR-125 asphalt and its final air-blown product, that is of AB-25 asphalt as well as of asphaltenes derived from both asphalts, are shown in Figure 7. The results show an apparent increase in molecular size of asphalt components upon air-blowing. It can also be seen from Figure 7 that the width of the apparent molecular size distribution increases markedly with the degree of asphalt oxidation. Thus, the results show that air-blowing of the asphalt leads not only to the increase in the asphaltene content (*see* Figure 6) but also to the significant increase in their apparent molecular size.

However, is this a result of condensation reactions leading to the formation of larger molecules, as is suggested by some authors (*20, 21*), or is this a result of intermolecular associations? To answer these questions, FIMS was used to investigate molecular weight distribution of both asphalts (SR-125 and AB-25) and their asphaltene fractions. Molecular weight envelopes of the straight-reduced SR-125 asphalt and its air-blown product, AB-25 asphalt, are shown in Figure 8. Both asphalts were 100% volatile under conditions of analysis. The first impression that emerges from examination of these spectra is the similarity of the molecular weight distributions for both samples: about 300–1800. The shape of the molecular weight envelope has changed slightly upon air-blowing of the asphalt. This may be attributable to compositional changes occurring in the asphalt upon oxidation, resulting in the formation of compound types of different sensitivities. The important result of this analysis is that no apparent increase in molecular weight was found. Molecular weight envelopes of asphaltenes derived from both asphalts (Figure 9) confirm this observation. The molecular weight distribution for both asphaltene samples is about 300–1800 and is the same as in the case of asphalts from which the asphaltenes were derived. The volatility of asphaltene (SR-125 and AB-25) samples under conditions of analysis was 92% and 83%, respectively. Comparison of the shape of the molecular weight envelopes of asphalts and of asphaltenes (Figures 8 and 9) shows that maxima of asphaltene curves are shifted towards lower molecular weight values. This implies that asphaltenes

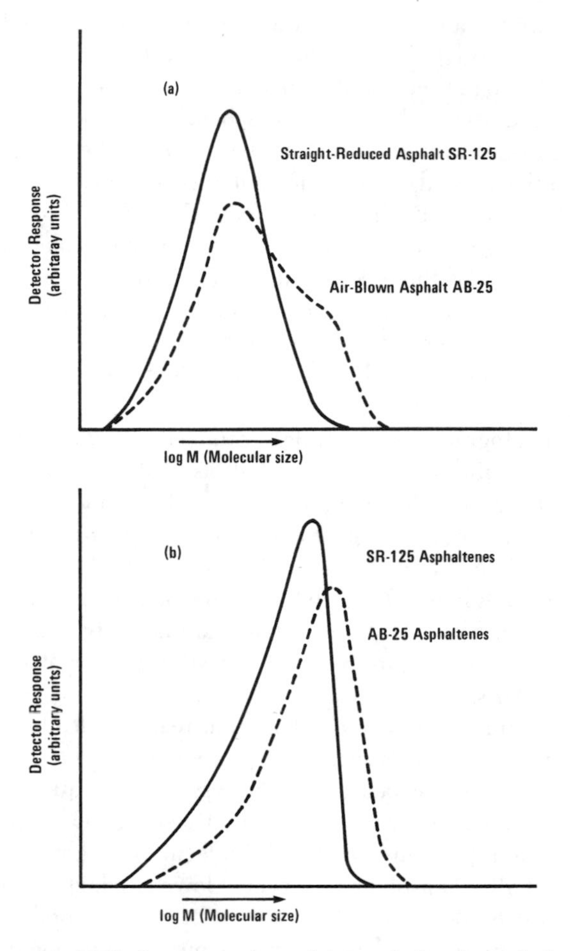

Figure 7. GPC chromatograms of: (a) asphalts; (b) asphaltenes

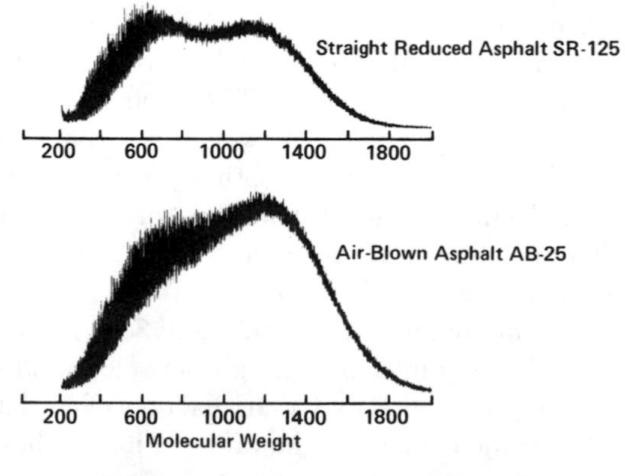

Figure 8. FIMS molecular weight profiles of asphalts

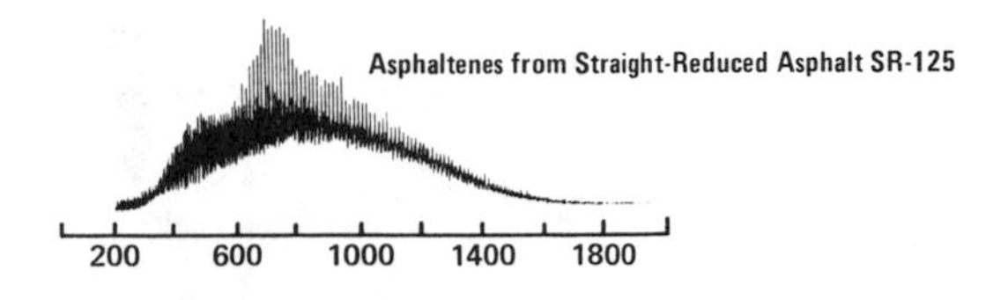

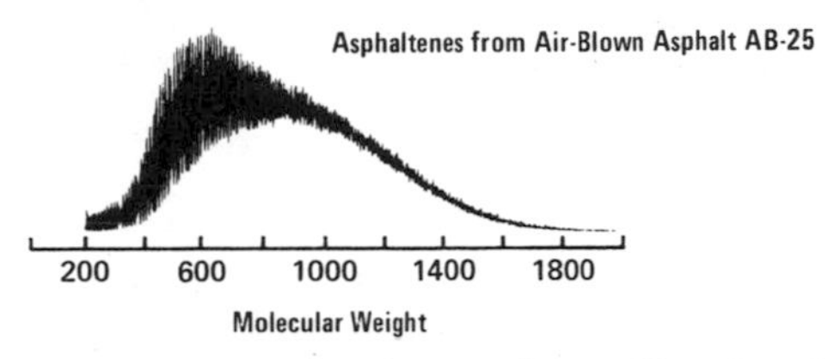

Figure 9. FIMS molecular weight profiles of asphaltenes

comprise compounds of relatively low (compared with other asphalt components) molecular weight but high polarity, resulting in strong intermolecular associations. Strong intermolecular association is probably responsible for the low volatility and limited solubility of compounds comprising asphaltenes.

Compositional Changes in Asphalt upon Air-Blowing. The chemical method of separation used in this study that utilizes ion exchange resins separates the asphalt into acids, bases, and a neutral (unreactive with ion exchange resins) fraction, which is further separated on alumina into three aromatic fractions and saturate hydrocarbon fraction (Figure 3). Thus, precipitation of asphaltenes is avoided. This procedure defines acids and bases in terms of their ability to associate with the functional group of the anion [—$N(CH_3)_3OH$] or the cation [—SO_3H] resin, respectively. Because of the polyfunctional compounds in the asphalt and the ability of some functional groups to associate with either resin, the separation sequence influences the compositions of the acid and base fractions. The fractions isolated from the asphalt as acids and bases comprise polar compounds capable of association through hydrogen bonding and possibly other molecular association forces.

The chemical separation method, together with the precipitation method, was used to investigate compositional changes occurring in the asphalt upon air-blowing. Relationships among the amounts of each compound-class fraction generated by the chemical separation method, the molecular size of compounds in each fraction, and the asphaltene content in the asphalt are shown in Figures 10 through 21. The increasing percentage of asphaltenes on the abscissa corresponds to their increasing concentration in asphalt during the air-blowing process. The molecular size (M_w) value on the ordinate was determined by GPC and is expressed in arbitrary units. It was shown earlier by FIMS that there is no apparent change in the molecular weight of asphalt components upon air-blowing. Thus, the increasing GPC molecular size values on the ordinate must indicate an increasing effect of intermolecular associations of compounds comprising a compound-class fraction.

Figure 10. Relationship between asphaltene and saturate contents (25)

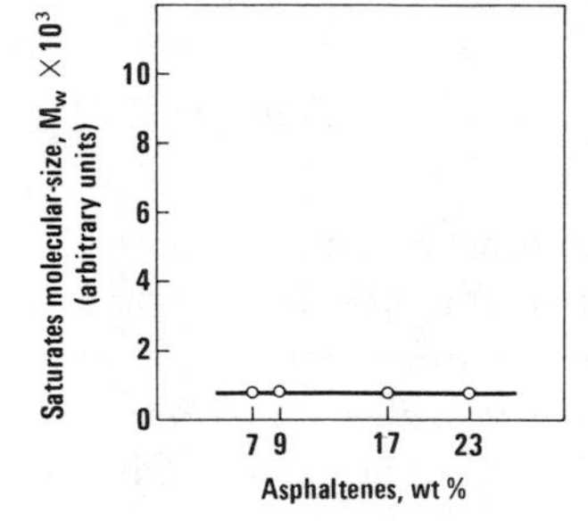

Figure 11. Relationship between asphaltenes content and molecular size of saturates (25)

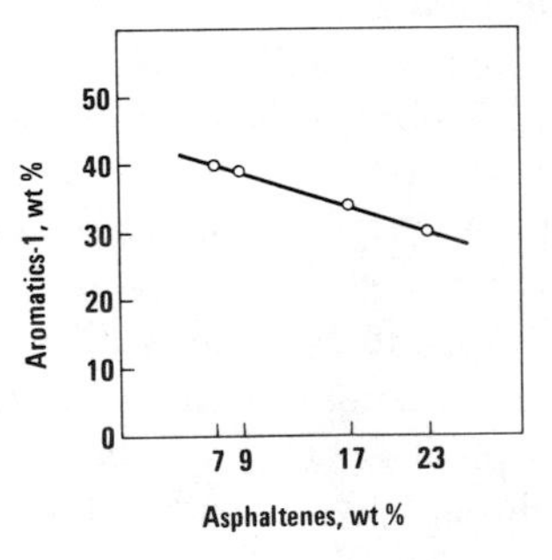

Figure 12. Relationship between asphaltene and aromatics-1 contents (25)

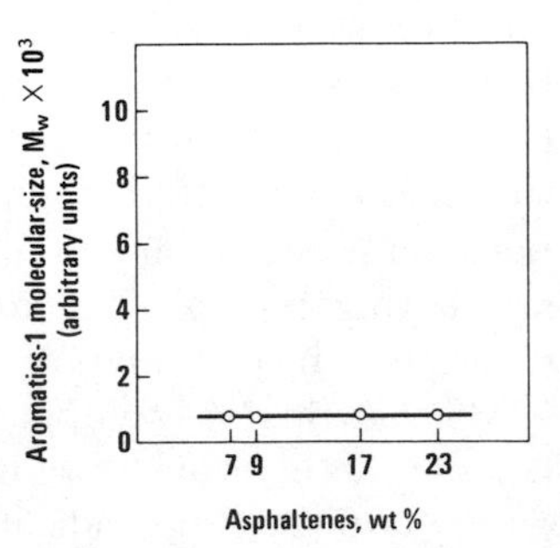

Figure 13. Relationship between asphaltenes content and molecular size of aromatics-1 (25)

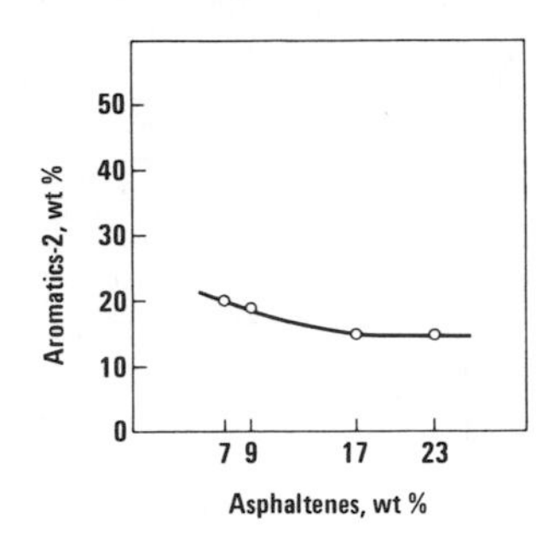

Figure 14. Relationship between asphaltene and aromatics-2 contents (25)

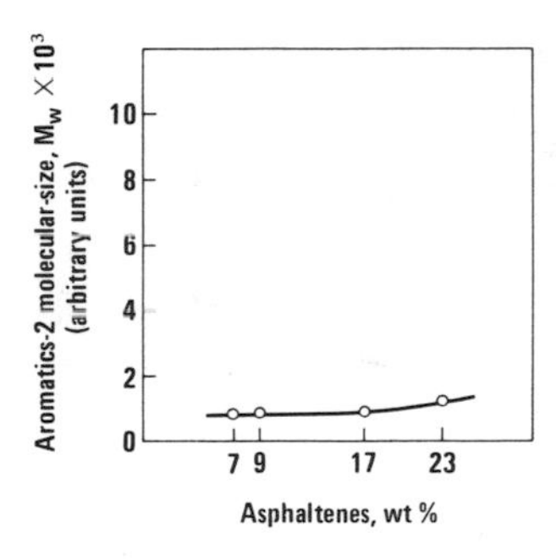

Figure 15. Relationship between asphaltenes content and molecular size of aromatics-2 (25)

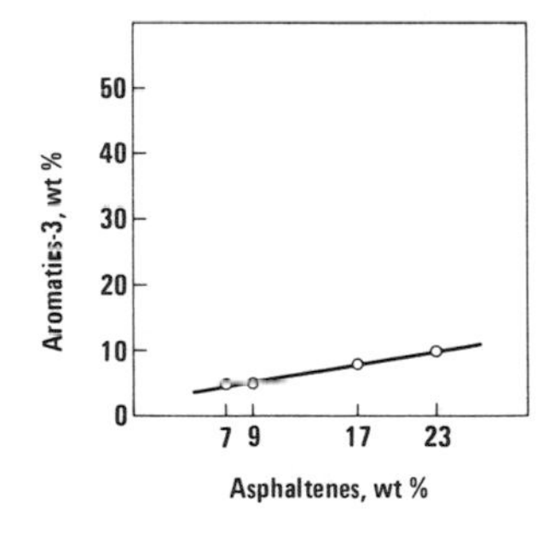

Figure 16. Relationship between asphaltene and aromatics-3 contents (25)

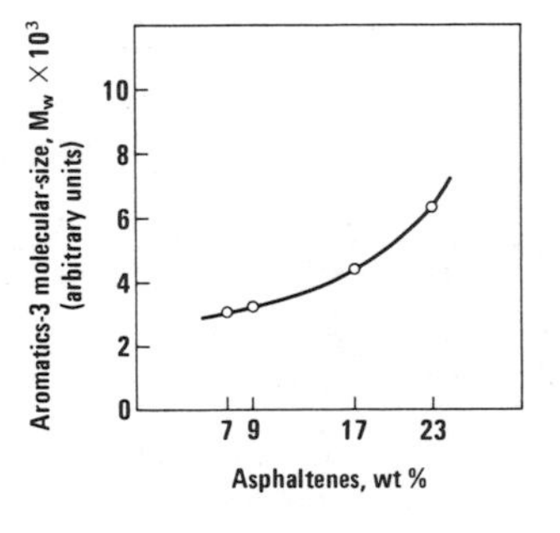

Figure 17. Relationship between asphaltenes content and molecular size of aromatics-3 (25)

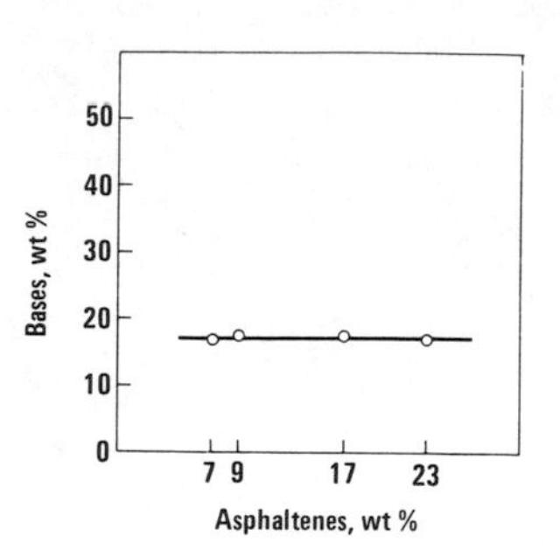

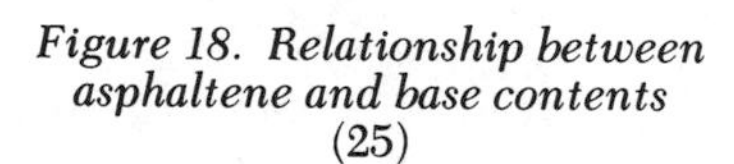
Figure 18. Relationship between asphaltene and base contents (25)

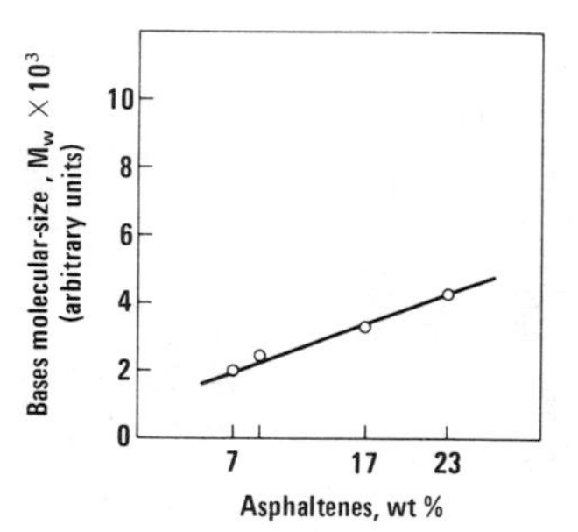

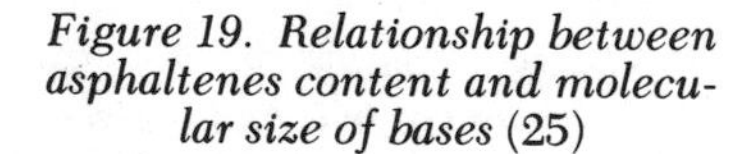
Figure 19. Relationship between asphaltenes content and molecular size of bases (25)

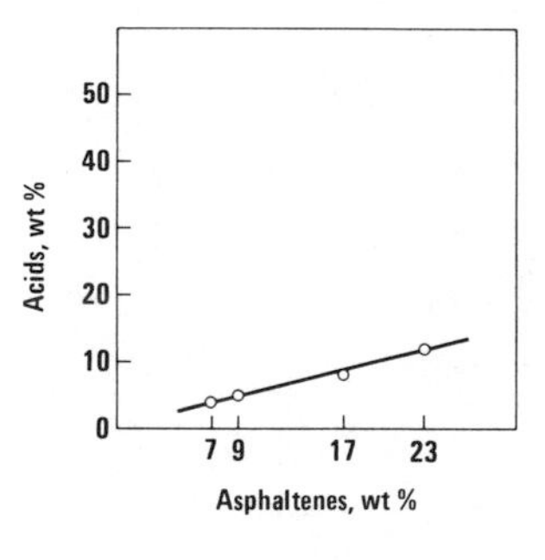

Figure 20. Relationship between asphaltene and acid contents (25)

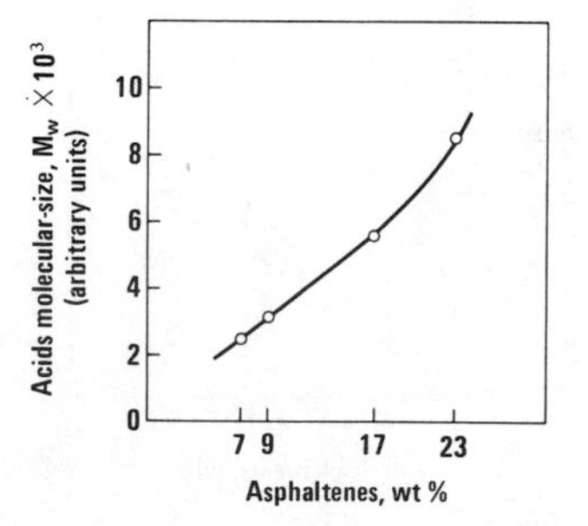

Figure 21. Relationship between asphaltenes content and molecular size of acids (25)

Figures 10 and 11 show that there is no change in the amount or in the apparent molecular size of saturates as a result of air-blowing. A significant decrease in the amount of the aromatics-1 with no change in apparent molecular size of components is observed in Figures 12 and 13. The results presented in Figures 14 and 15 show that the amount of the aromatics-2 fraction decreases upon air-blowing of the asphalt and that the apparent molecular size of components of this fraction increases. Air-blowing increases the amount of the aromatics-3 fraction (Figure 16), with a simultaneous significant increase in apparent molecular size of the components present in this fraction (Figure 17). Compositional changes within the aromatic fractions of the asphalt can be interpreted as the formation of more polar or more strongly associating compounds (aromatics-3 and also acids, as will be shown later) at the expense of relatively nonpolar, weakly associating compounds (aromatics-1 and aromatics-2). The specific nature of functionalities formed during air-blowing is unknown, but it is suggested that these functionalities contribute to intermolecular associations through such association forces as hydrogen bonding and permanent or induced dipoles, resulting in high apparent molecular size values. Yen (*24*) has suggested that large π-aromatic systems are capable of electron donor–electron acceptor activity, resulting in intermolecular associations. It is possible that dehydrogenation takes place, as suggested by Corbett (*20*) and Haley (*21*), and condensed aromatic ring systems are formed.

Data in Figure 18 show that the amount of bases remains constant upon air-blowing of the asphalt. However, Figure 19 shows that the apparent molecular size of the bases increases significantly with an increase in asphalt air-blowing. This result may be attributable to an increase in hydrogen-bonding basicity of that fraction. The formation of hydrogen-bonding bases by oxidation of asphalts at elevated temperatures has been suggested by Barbour and Petersen (*17*). Sulfur-containing molecules are also proposed by these authors as important hydrogen-bonding bases in asphalt.

Data in Figure 20 show that air blowing of the asphalt increases the amount of acids. These results imply that acids are being formed at the expense of the aromatic fractions, particularly aromatics-1 and/or aromatics-2. The dramatic increase in apparent molecular size of acids is illustrated in Figure 21. A variety of acidic functionalities is known to form association complexes (*15*, *16*).

Compositional changes occurring in the asphalt upon air-blowing and resulting in the formation of asphaltenes can be summarized as the formation of polar species capable of strong intermolecular associations, the latter being responsible for apparent high molecular size values observed by GPC. No increase in molecular weight of asphalts was observed, as indicated by FIMS. The increased polarity or polarizability of the asphalt upon air-blowing results in formation of agglomerates of limited solubility. Treatment of an asphalt with normal alkane (e.g., heptane) upsets the solubility equilibrium of the very complex mixture of asphalt components; the more polar or polarizable compounds capable of forming strong intermolecular associations precipitate as asphaltenes.

Conclusions

1. Asphaltenes in the straight-reduced asphalt are formed by intermolecular associations of naturally occurring hydrogen-bonding acids and bases.
2. The increase in asphaltenes in air-blown asphalts results from chemical reactions occurring in the asphalt upon oxidation, resulting in the formation of molecular aggregates.
3. The apparent increase in molecular size (GPC) of asphalt components on air-blowing results primarily from increased intermolecular associations and not from actual increase in molecular weight.
4. Asphaltenes appear to be agglomerates of polar compounds; these agglomerates, formed through intermolecular associations, vary in composition because of the chemically nonspecific separation procedure, which is based only upon solubility parameters.

Acknowledgment

Appreciation is expressed to S. E. Buttrill, Jr., of SRI International, for his assistance in providing field ionization mass spectrometry data. Appreciation is also expressed to D. R. Latham, J. F. McKay and J. C. Petersen of the Laramie Energy Technology Center for helpful suggestions and review of this manuscript. The assistance of the Laramie Energy Technology Center for partial support of this research is gratefully acknowledged.

Literature Cited

1. Hoiberg, A. J. "Bituminous Materials: Asphalts, Tars and Pitches"; Interscience: New York, 1965; Vol. 2, p. 157.
2. Long, R. B. *Am. Chem. Soc., Div. Pet. Chem., Prepr.* (Washington, D.C., Sept. 1979) *24*(4), 891.
3. Boduszynski, M.; Chadha, B. R.; Pineles, H. *Fuel* **1977,** *56,* 145.
4. Boduszynski, M.; Szkuta-Pochopien, T. *Fuel* **1977,** *56,* 149.
5. Boduszynski, M.; Chadha, B. R.; Szkuta-Pochopien, T. *Fuel* **1977,** *56,* 432.
6. Boduszynski, M.; McKay, J. F.; Latham, D. R. *Proc. Assoc. Asphalt Paving Technol. Louisville, Kentucky,* **1980,** *49,* 123–143.
7. Jewell, D. M.; Weber, J. H.; Bunger, J. W.; Plancher, H.; Latham, D. R. *Anal. Chem.* **1972,** *44,* 1391.
8. Boduszynski, M.; Warzechowa, L., *Nafta* (Katowice, Poland)**1976,** *32,* 198.
9. Corbett, L. W. *Anal. Chem.* **1969,** *41,* 576.
10. St. John, G. A.; Buttrill, S. E. Jr.; Anbar, M. In "Organic Chemistry of Coal," *ACS Symp. Ser.* **1978,** *71,* 223.
11. McKay, J. F.; Amend, P. J.; Cogswell, T. E.; Harnsberger, P. M.; Erickson, R. B.; Latham, D. R. In "Analytical Chemistry of Liquid Fuel Sources, Tar Sands, Oil Shale, Coal and Petroleum," *Adv. Chem. Ser.* **1978,** *170,* 128.

12. Speight, J. G.; Moschopedis, S. E. *Am. Chem. Soc., Div. Pet. Chem., Prepr.* (Washington, D.C., Sept., 1979) *24*(4), 910.
13. Moschopedis, S. E.; Fryer, J. F.; Speight, J. G. *Fuel* **1976,** *55,* 227.
14. Petersen, J. C. *Fuel* **1967,** *46,* 295.
15. Petersen, J. C. *J. Phys. Chem.* **1971,** *75,* 1129.
16. Petersen, J. C.; Barbour, R. V.; Dorrence, S. M.; Barbour, F. A.; Helm, R. V. *Anal. Chem.* **1971,** *43,* 1491.
17. Barbour, R. V.; Petersen, J. C. *Anal. Chem.* **1974,** *46,* 273.
18. Hoiberg, A. J. "Bituminous Materials: Asphalts, Tars and Pitches"; Interscience: New York, 1965; Vol. 2, pp. 257–260.
19. Corbett, L. W.; Swarbrick, R. E. *Proc. Assoc. Asphalt Paving Technol.* **1960,** *29,* 104.
20. Corbett, L. W. *Ind. Eng. Chem. Process Des. Dev.* **1975,** *14,* 181.
21. Haley, G. A. *Anal. Chem.* **1975,** *47,* 2432.
22. Brule, B., J. *Liquid Chromatogr.* **1979,** *1*(2), 165.
23. Such, C.; Brule, B.; Baluja-Santos, C., J. *Liquid Chromatogr.* **1979,** *2*(3) 437.
24. Yen, T. F. *Am. Chem. Soc., Div. Fuel Chem., Prepr.* (Los Angeles, Mar.–April, 1971) *15*(1) 93.
25. Boduszynski, M. M. *Am. Chem. Soc., Div. Pet. Chem., Prepr.* (Washington, D.C., Sept., 1979) *24*(4), 935.

RECEIVED June 23, 1980.

8

Size Characterization of Petroleum Asphaltenes and Maltenes

G. HALL and S. P. HERRON

Mobil Research and Development Corporation, Paulsboro, NJ 08066

The molecular size distributions and the size-distribution profiles for the nickel-, vanadium-, and sulfur-containing molecules in the asphaltenes and maltenes from six petroleum residua were determined using analytical and preparative scale gel permeation chromatography (GPC). The size distribution data were useful in understanding several aspects of residuum processing. A comparison of the molecular size distributions to the pore-size distribution of a small-pore desulfurization catalyst showed the importance of the catalyst pore size in efficient residuum desulfurization. In addition, differences between size distributions of the sulfur- and metal-containing molecules for the residua examined helped to explain reported variations in demetallation and desulfurization selectivities. Finally, the GPC technique also was used to monitor effects of both thermal and catalytic processing on the asphaltene size distributions.

Recently, petroleum residua have been studied extensively (*1, 2*) because of the increasing importance of heavier fuels. Both the asphaltene (pentane-insoluble) and maltene (pentane-soluble) components of residua are of interest, and since their properties overlap, a complete study of petroleum residua must consider both asphaltenes and maltenes. One area that has received considerable attention has been the size characterization of asphaltenes and maltenes (*3, 4, 5*). Size distribution data are useful both in understanding the fundamental chemistry of asphaltenes and maltenes and in observing the effects of various processes on residua sizes.

In this study, six petroleum residua were characterized by a combination of preparative- and analytical-scale gel permeation chromatography (GPC). Each residuum was separated initially by pentane deasphalting into an asphaltene and maltene pair, both of which were separated further by

0065-2393/81/0195-0137$05.00/0

preparative GPC on Styragel. The apparent molecular size distributions were obtained on an analytical GPC system using μ-Styragel columns. These distributions, with the sulfur, nickel, and vanadium measurements for each cut, were used to obtain size distribution profiles for the sulfur-, nickel-, and vanadium-containing molecules. The molecular and elemental size distribution data were compared with the pore size distribution data of a small-pore desulfurization catalyst to illustrate the importance of catalyst pore size for efficient desulfurization and demetallation. In addition, the effects of both thermal and catalytic processing on asphaltene size distributions were monitored using these data.

Experimental

Samples. Table I lists the six residua studied and their sulfur, nickel, vanadium, and weight percent asphaltenes data. The Arabian Light is a vacuum (1000 + °F) residuum, while the other five are atmospheric (650 + °F) residua. The samples were analyzed as received from the refinery distillation tower.

Sample Preparation. The residua samples were separated into asphaltenes and maltenes by deasphalting the resid with a 25:1 (v/v) amount of n-pentane. After stirring, the mixture was allowed to sit overnight, then filtered through a 0.45-μ porous glass filter. The asphaltenes were washed with several portions of pentane and dried under vacuum at 90°C. Pentane was evaporated from the filtrate to yield the maltenes.

Preparative GPC. The preparative GPC work was performed on the experimental setup shown in Figure 1. Four 1-in. i.d. glass columns were packed with Styragel (Waters Associates) with 1 ft 10^4 Å porosity, 2 ft 500 Å porosity, and 1 ft 100 Å porosity. The Styragel porosites were chosen to give good resolution for the entire range of molecular sizes found in residua. Two separate column systems were used—one for maltenes, the other for asphaltenes.

Tetrahydrofuran (Burdick and Jackson "distilled in glass") and pyridine (Baker Instra-Analyzed) were the mobile phases. All of the maltene sample was eluted by tetrahydrofuran; however, pyridine was required to remove a

Table I. Petroleum Residua Analyzed

Description	*S (%)*	*Ni (ppm)*	*V (ppm)*	*Weight Percent Asphaltenes*
Aramco	2.77	5	22	3.6
Arabian Light vacuum	4.17	17	80	13.2
Kuwait	4.24	13	50	6.7
Lagomedio	1.80	18	204	5.0
Prudhoe Bay	1.61	17	49	4.8
Wilmington	1.96	93	67	7.8

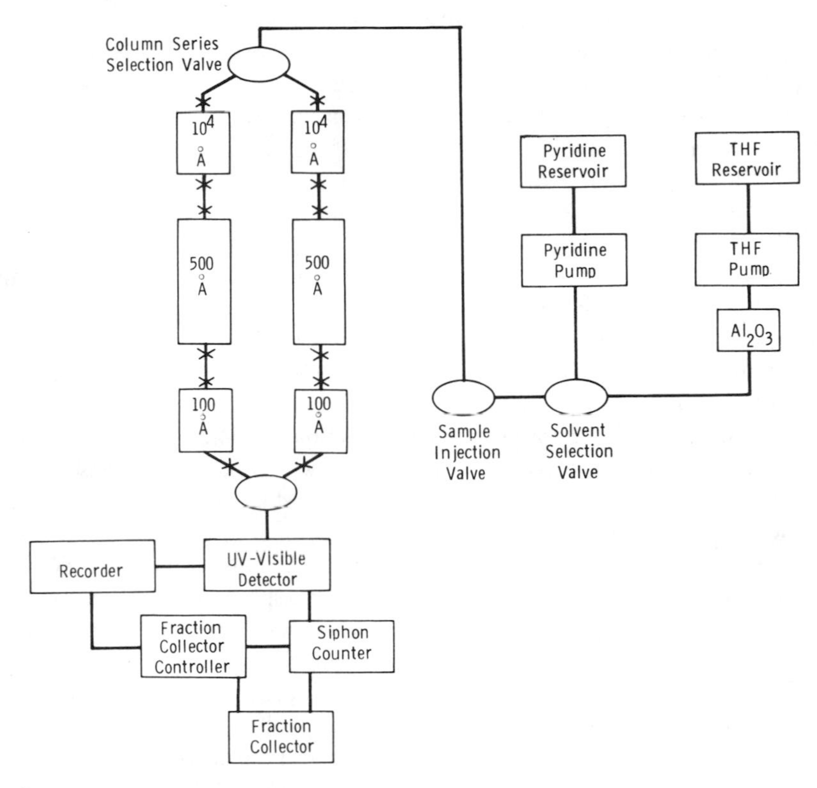

Figure 1. Experimental setup for preparative GPC work

portion of the asphaltene sample (5–10%) not eluted by tetrahydrofuran. A Milton–Roy reciprocating pump with stainless steel 316 and Teflon surfaces was used for tetrahydrofuran. Since pyridine leaches nickel from stainless steel, a Fluidmetering positive displacement piston pump with Teflon and ceramic surfaces was used for pyridine. A 1-ft × ⅜-in. o.d. Al_2O_3 (Alcoa F20) column was inserted between the THF pump and the GPC column to remove any peroxides from the tetrahydrofuran.

A 100-mg sample was loaded into a 4-mL glass sample loop and tetrahydrofuran was added to solubilize the sample before the solution was flushed onto the column. The flow rate was 10 mL/min. Four to eight passes were made for each sample to obtain sufficient material for all subsequent analyses.

The column eluate was monitored with an Isco dual beam UV–vis detector operating at 310 nm and 546 nm. Ten-milliliter fractions were collected with an Isco fraction collector triggered by a siphon counter, and after completion of a run, adjacent 10-mL fractions were combined into five cuts.

The fractions were evaporated to near dryness with a roto-evaporator, then transferred to a small, tared beaker for further evaporation under N_2 on a hot plate. Final evaporation was accomplished in a vacuum oven at 90°C.

ELEMENTAL ANALYSES. Sulfur measurements for preparative cuts and raw and control samples were made on 15–50-mg samples using ASTM Method D1552. The precision was ± 10% relative for maltenes and ± 5% relative for asphaltenes as determined by multiple measurements on several asphaltenes and maltenes.

Flame atomic absorption determinations for both preparative cuts and raw and control samples of nickel and vanadium were made on 10–150-mg samples. The precision was ± 10% relative, as determined by replicate measurements on a series of samples, except for very low metal level samples (<10 ppm), where the precision was ± 20%.

PREPARATIVE GPC RESULTS. The sulfur, nickel, vanadium, and weight data for the prep runs are given in Table II. These raw data were subsequently corrected for solvent residues. Although the highest grade solvents were used, residue weights and metal contents must be considered because of the large amounts of solvents evaporated to dryness and the small sample sizes involved. For all but the pyridine-containing cut for the asphaltenes, the corrections were negligible (<2%). However, for the pyridine cut, the corrections to the weights and metal concentrations were on the order of 10%.

Table III lists the material balances for the preparative separations. These are the percent weight recoveries for either asphaltene or maltene defined, using the sulfur balance for an example, as the sum of the amount of sulfur in each cut times the cut weight percent divided by the total sulfur. In general, the balances are in the 80–120% range, which is reasonable considering the amount of sample handling involved. The recoveries are out of line only in a few cases, most notably the Prudhoe Bay maltene nickel balance. In addition, a comparison of the calculated elemental values for the total residua differ somewhat from the raw total values for several residua. These discrepancies are probably attributable to the small samples, multiple sample manipulations, and compounding of individual errors when the asphaltene and maltene data are summed. The data-fitting routine described in the next section was used to obtain a set of best fit data, which were used in the subsequent size calculations.

TREATMENT OF RAW DATA. The method of Lagrange multipliers was used to normalize the experimental weight, sulfur, nickel, and vanadium data. A best fit of the experimental data subject to the constraints of 100% recoveries was obtained, using weighting factors determined by the analytical precision of each measurement.

Analytical GPC. Size measurements were made on an analytical GPC system using ⅜-in. o.d. μ-Styragel columns (Waters Associates) with 1 ft 10^4 Å porosity, 2 ft 500 Å porosity, and 1 ft 100 Å porosity. The samples were run as 0.1% tetrahydrofuran solutions prefiltered through a 0.5-μ Fluoropore filter (Millipore Corporation) with THF at 2 mL/min as the mobile phase.

Four primary problems must be considered in performing GPC studies of asphaltenes and maltenes. These are poor weight recoveries from GPC columns through sample adsorption, nonuniform detector responses for different sample sizes and/or types, uncertainty in selection of appropriate column calibrations, and non-size-exclusion behavior because of abnormal aromatics retention on Styragel. All these problems will be addressed in the following sections.

Tetrahydrofuran was used since it is a good solvent for residua samples. However, because tetrahydrofuran did not elute 100% of the asphaltene samples from the prep columns, the sample recovery from the analytical columns was checked using the asphaltenes from an Arabian Light vacuum residuum. After sample injection onto the μ-Styragel columns, the effluent was collected and concentrated. UV absorbances at several wavelengths (254–546 nm) were measured and compared with that expected for a solution with no column losses. The recovery for duplicate runs was about 90%, and was the same as for the preparative column. However, the material eluted only with pyridine on the preparative column could be chromatographed on the analytical column, so the losses were spread out over all five cuts. Several other solvent systems (chloroform, 5% methanol–chloroform (*3*), 10% 1,2,4-trichlorobenzene–chloroform (*6*), pyridine) were tried, but tetrahydrofuran gave the best results. The vacuum residuum asphaltenes gave the poorest recovery (90%) and other samples—atmospheric residua asphaltenes and maltenes—had better recoveries. Therefore, the analytical GPC was considered an acceptable method.

An autosampler (Varian) with a 50-μL sample loop injected the sample through a 2-μm, in-line filter onto the column.

Peak height measurements were made with a UV detector (Waters Associates) at 254 nm. Response factors for different sized molecules were determined by measuring the UV absorbance for the five cuts obtained from the preparative GPC of the asphaltenes from an Arabian Light vacuum residuum. A variation in extinction coefficients from 34.5 to 61.5 L/(g – cm)$^{-1}$ was found for the cuts. A corrected total curve was obtained by summing the size distributions of the individual cuts, weighted by their weight fractions and response factors (i.e., relative extinction coefficients). The mean sizes for the corrected and uncorrected total curves differed by less than experimental error. However, an examination of response factors shows that only Cut 5 (<10%) has a greatly different factor and it contains the smallest molecules, which are not as important in the mean size determination as the larger molecules. Therefore, a variation in response factors does not introduce any significant error into the mean size determination.

Data acquisition was accomplished with the aid of a Varian Aerograph Chromatography Data System 220i. The UV detector was connected to the Varian system, which integrated the peak-height–retention-time data. The integral area vs. elution time distribution was converted to the desired differential distribution.

Table II. GPC Prep Data

	Asphaltene						*Maltene*					
	Raw Sample	*Cuts* 1	2	3	4	5	*Raw Sample*	*Cuts* 1	2	3	4	5
Aramco												
sulfur (%)	6.58	5.17	6.50	5.31	5.60	4.60	3.12	3.90	2.40	2.20	2.90	3.60
nickel (ppm)	63	43	41	38	60	193	13	12	5	8	3	13
vanadium (ppm)	353	315	251	212	279	388	16	29	13	10	9	20
weight (%)		32.1	23.4	24.0	12.9	14.7		17.3	27.6	16.0	28.5	15.6
Arabian Light vacuum												
sulfur (%)	6.31	5.54	5.85	6.02	4.17	5.31	3.87	34.0	2.27	2.86	3.55	6.00
nickel (ppm)	116	152	101	69	52	324	8	18	4	6	3	20
vanadium (ppm)	349	401	328	238	251	397	26	56	13	12	9	22
weight (%)		12.8	27.0	33.4	25.0	9.4		20.3	25.2	14.7	22.2	16.1
Kuwait												
sulfur (%)	7.43	7.23	7.28	7.07	4.76	5.04	3.82	4.70	3.31	3.18	5.65	3.68
nickel (ppm)	122	172	128	66	110	220	4	3	1	2	1	166
vanadium (ppm)	444	520	404	317	370	483	25	11	19	13	35	378
weight (%)		14.2	28.6	35.6	18.1	7.9		10.0	22.8	41.4	23.9	2.1

Table II. GPC Prep Data (*continued*)

	Asphaltene						*Maltene*					
		Cuts						*Cuts*				
	Raw Sample	*1*	2	3	4	5	*Raw Sample*	*1*	2	3	4	5
Lagomedio												
sulfur (%)	5.10	4.45	5.19	3.95	4.18	3.45	2.20	2.13	1.67	1.43	1.64	2.87
nickel (ppm)	260	296	195	145	214	370	15	40	16	12	5	13
vanadium (ppm)	2570	2448	1921	1753	2457	2139	105	266	75	60	65	127
weight (%)		29.0	24.2	19.2	15.9	9.4		13.9	22.5	15.3	25.6	16.2
Prudhoe Bay												
sulfur (%)	2.84	3.02	2.97	2.65	2.22	1.97	2.17	3.06	2.03	2.59	2.21	2.54
nickel (ppm)	170	213	195	125	188	592	13	94	28	9	9	19
vanadium (ppm)	335	339	350	275	335	347	24	50	21	18	18	27
weight (%)		10.8	28.7	43.7	20.3	10.0		17.6	19.7	13.7	25.1	25.3
Wilmington												
sulfur (%)	2.50	2.50	2.35	1.80	1.80	1.60	1.99	2.53	1.92	1.80	1.73	1.99
nickel (ppm)	550	420	340	415	880	455	53	94	45	38	35	53
vanadium (ppm)	480	325	320	560	1040	300	32	56	28	22	46	32
weight (%)		30.9	24.0	18.8	12.6	12.6		12.8	20.8	24.0	24.0	12.1

Table III. Material Balances (%)

	Weight	*Sulfur*	*Nickel*	*Vanadium*
Aramco				
asphaltene	107.2	89.0	109.1	86.1
maltene	105.0	108.6	58.0	99.6
Arabian Light vacuum				
asphaltene	107.6	92.6	97.5	91.6
maltene	98.4	88.7	124.0	84.6
Kuwait				
asphaltene	104.3	92.6	99.8	91.8
maltene	100.1	103.8	111.6	108.3
Lagomedio				
asphaltene	97.7	83.8	88.4	81.9
maltene	93.4	80.7	96.5	95.4
Prudhoe Bay				
asphaltene	113.5	105.7	135.8	107.4
maltene	101.4	114.7	234.3	111.8
Wilmington				
asphaltene	98.8	84.1	85.6	96.0
maltene	93.7	92.0	100.2	108.0

SIZE CALIBRATION. The μ-Styragel columns were calibrated to transform the elution data from the time domain to the size domain using both n-alkane and polystyrene (Pressure Chemicals) standards. The n-alkane sizes are related to the carbon number by Equation 1 (*7*).

$$\text{Size (Å)} = (1.25 \times \text{carbon number}) + 2.5 \tag{1}$$

The polystyrene sizes were not the sizes commonly assumed for polystyrenes based on extended chains, but instead were determined from the viscosities (measured by Arro Labs, Joliet, Illinois) as described in Reference *8*. Other types of standards and methods of calibration have been used for the GPC of residua samples (*3, 9*), including the Benoit universal calibration (*10*). However, these all relate the molecular weight to elution time, and for this work a molecular-size–elution-time relationship was needed. The polystyrene and n-alkane sizes were used to construct a ln size vs. elution time calibration that was fit to a fourth-order polynomial to give a smooth curve.

This calibration does not assume that the n-alkanes and polystyrenes are typical of residual molecules. However, they do provide well-defined size standards in the elution time range of interest. No assumptions can be made concerning the shapes of the asphaltene or maltene molecules. Therefore, the GPC size calculated is defined as the critical molecular dimension, which determines if the asphaltene or maltene molecule will diffuse into the pores of the GPC packing. This size is assumed to be related to the size parameter that determines the molecular diffusion into hydrotreating catalyst pores.

TIME TO SIZE TRANSFORMATION. The differential elution time distribution was normalized and then converted to the size distribution using the calibration curve described above. Since the time–size relationship was nonlinear, it was necessary to use the transformation given in Equation 2 (*11*),

$$g(s) = \frac{f(t)}{\frac{ds}{dt}} \tag{2}$$

where $g(s)$ and $f(t)$ are the distributions in the size and time domains, respectively, and ds/dt is the derivative of the calibration equation. A mean size $\bar{s}$ was determined for the size distribution according to Equation 3.

$$\bar{s} = \int s g(s) ds \tag{3}$$

The precision for the mean size measurements was ± 5%.

The size distributions of the sulfur- and metal-containing molecules were determined by summing the individual cut size distributions weighted by the fraction of total sulfur (or metal) in that cut. Equation 4 defines the mean size for the distribution where $\bar{x}$ is sulfur, nickel, or vanadium.

$$\bar{x} = \sum_i \frac{w_i x_i size_i}{W_T X_T} \tag{4}$$

w_i = cut weight
x_i = cut sulfur, nickel, or vanadium
$size_i$ = cut mean size
W_T = total sample weight
X_T = total sample sulfur, nickel, or vanadium

Previously, researchers have noted that the GPC of asphaltenes on Styragel with tetrahydrofuran as the mobile phase show tailing indicative of delayed elution (*12*). Asphaltenes are large molecules made up in part of condensed polynuclear aromatics (PNAs) (*1*) that are not eluted strictly on a size basis in tetrahydrofuran, but instead have abnormally long retention times. This is thought to be attributable to the interaction of the pi system of the PNAs with the highly aromatic Styragel, which is a styrene–divinylbenzene copolymer (*13*). The abnormal behavior should not be as significant for asphaltenes as for pure PNAs, since the condensed aromatics in asphaltenes have alkyl sidechains, fused naphthenic rings, etc., so any pi interactions will compete with normal size exclusion behavior. Several condensed PNAs were incorporated into a trial calibration curve to test this effect. The asphaltene sizes determined with this new curve were not significantly different from those determined with the *n*-alkane–polystyrene curve. Therefore, the original *n*-alkane–polystyrene calibration curve was used.

Discussion of Method

The sizes determined in this work are the apparent molecular sizes and not necessarily the sizes of the asphaltene and maltene molecules at process conditions. Association efforts for asphaltene molecules have been observed for both vapor-phase osmometry molecular weight and viscosity measurements (*14*, *15*). The sizes reported here were measured at 0.1 wt % in tetrahydrofuran at room temperature. Other solvent systems (chloroform, 5% methanol–chloroform, and 10% trichlorobenzene–chloroform) gave similar size distributions. Under these conditions, association effects should be minimized but may still be present. At process conditions (650–850°F and 5–30% asphaltene concentration in a maltene solvent), the asphaltene sizes may be smaller. However, for this work the apparent sizes determined can be meaningfully correlated with catalyst pore size distributions to give reasonable explanations of the observed differences in asphaltene and maltene processabilities (vide infra). In addition, the relative size distributions of the six residua are useful in explaining the different processing severities required for the various stocks. Therefore, the apparent sizes determined here have some physical significance and will be referred to just as sizes.

Figure 2 shows the size distributions for a typical GPC prep separation. A gradual decrease in mean sizes for Cuts 1–5 is observed, where Cut 1 is eluted first from the column and contains the largest molecules. As the figure shows, the preparative method gives a good separation of the sample into different molecular size fractions.

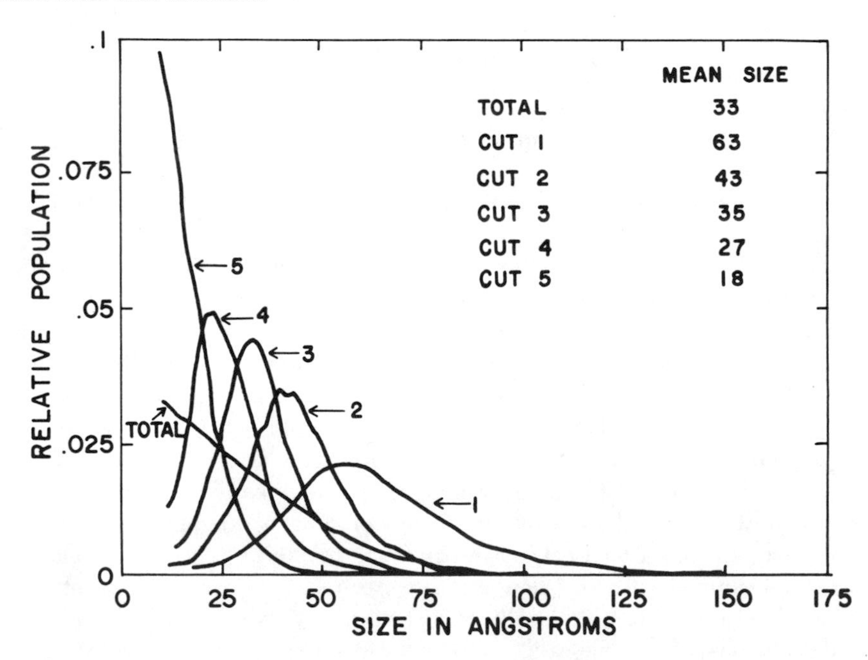

Figure 2. Size distribution of Aramco maltene

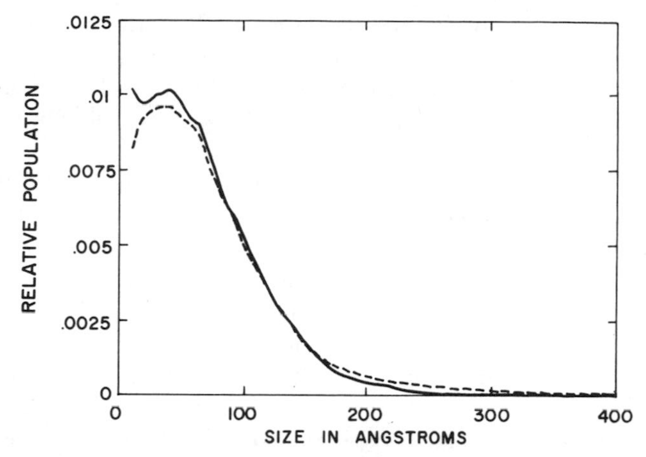

Figure 3. Comparison of experimental and best fit curves for Lagomedio asphaltene: (—) experimental, (---) best sum

A comparison of the experimentally determined size distribution for the Lagomedio asphaltenes and the best fit size distribution obtained by summing the size distributions of the individual cuts weighted by their weight fractions is shown in Figure 3. The two curves are quite similar, indicating the overall self-consistency of the data.

Table IV lists the overall mean sizes and the mean sizes for the sulfur-, nickel-, and vanadium-containing molecules obtained for duplicate determinations on an Arabian Light vacuum asphaltene performed two years apart. The precision of the measurements shows the reproducibility of this technique.

Results and Discussion

Both asphaltene and maltene molecular size distributions were compared with the pore size distribution of a small pore desulfurization catalyst. Figure 4 shows the Kuwait maltene and asphaltene size distributions along with the catalyst pore size distribution. Most of the maltene molecules are small enough to diffuse into the catalytic pores. In contrast, the Kuwait asphaltenes have a

Table IV. GPC Prep Method Reproducibility

Arabian Light Vacuum Asphaltene

Mean Sizes	*October 1977*	*August 1979*	*Percent Deviation*
Molecular	62.0	60.9	1.8
Sulfur	75.0	71.7	4.4
Nickel	80.8	82.9	2.6
Vanadium	77.2	77.8	0.8

substantial number of molecules too large to fit into the pores and be processed. Previous workers also have found this exclusion of asphaltenes from the catalyst (*16*, *17*). These data dramatically illustrate the importance for efficient processing in matching the catalyst pore size distribution to the molecular sizes.

Table V lists the molecular mean sizes and the mean sizes for the sulfur, nickel, and vanadium compounds of the residua studied. The sulfur and metal compounds have mean sizes larger than the overall molecular sizes. Drushel has postulated an asphaltene structure containing more than one sulfur atom per molecule (*16*). Since the sulfur and metals mean size calculations depend on thc fraction of the heteroatom, larger molecules containing more than one heteroatom will be weighted more heavily in the size determination. Therefore, the mean sizes for the sulfur- and metal-containing molecules in the asphaltenes would be expected to be greater than the mean molecular size calculated with each molecule weighted equally. Previous work in this laboratory also supports this expectation. Gradient elution chromatography on the Arabian Light residuum vacuum asphaltenes showed that the sulfur, nickel, and vanadium distributions differed from the overall weight distributions. Both sulfur and metals were found mainly in later eluting, larger molecular weight fractions (*18*).

Since residuum hydroprocessing involves both demetallation and desulfurization reactions, the residuum metal and sulfur-containing molecules and their sizes are important. As shown in Table V and illustrated in Figure 5 for the Arabian Light vacuum asphaltene, the sulfur, nickel, and vanadium compounds for the asphaltene from any given residuum have similar sizes;

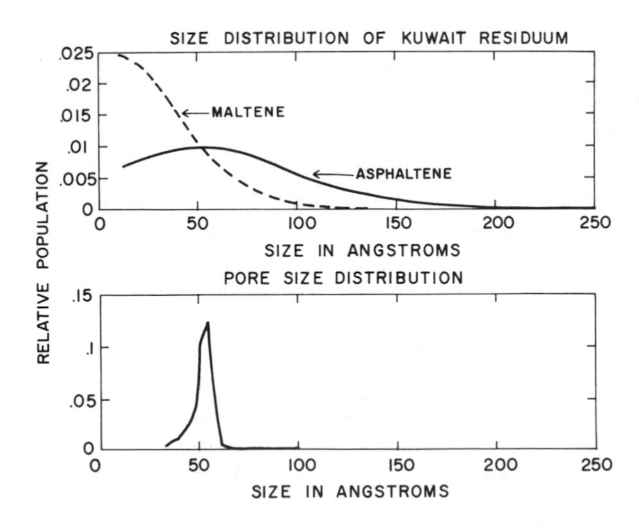

Figure 4. Comparison of small pore catalyst and Kuwait asphaltene and maltene

Table V. Mean Size Data

	Mean Size (Å)			
	Total	*Sulfur*	*Nickel*	*Vanadium*
Asphaltenes				
Kuwait	75	87	91	88
Arabian Light vacuum	62	75	81	77
Lagomedio	70	79	83	80
Aramco	65	80	77	84
Prudhoe Bay	75	92	92	91
Wilmington	62	74	66	62
Maltenes				
Kuwait	38	42	62	47
Arabian Light vacuum	43	46	54	63
Lagomedio	37	40	52	49
Aramco	33	36	40	41
Prudhoe Bay	44	54	71	61
Wilmington	33	39	39	38
Total				
Kuwait	40	45	64	50
Arabian Light vacuum	46	50	57	65
Lagomedio	39	42	54	51
Aramco	34	38	41	43
Prudhoe Bay	45	55	72	63
Wilmington	36	41	41	40

however, the sulfur-containing molecules are smaller than the metal-containing molecules. This behavior is seen for most residua, with the exceptions of Wilmington and Aramco residua. A combination of asphaltene and maltene data give the total residuum sizes, which show the same trend as the maltene data, that is, the average sizes of the sulfur compounds are significantly smaller than those of the metal compounds.

Catalyst selectivity differences have been found for sulfur and metals removal in residuum hydroprocessing (*19*). Mass transfer limitations are believed to be important (*20*). The data reported here show that the metal-containing molecules are larger; consequently, they should be more subject to diffusion restrictions than the sulfur-containing molecules. Therefore, it will be more difficult for a small pore catalyst to demetallate a residuum than to desulfurize it.

The GPC size data also were used to monitor the effect of various processes on the asphaltene molecular size distributions. Only the asphaltenes were studied in these experiments since they are the hardest to process, and they cause the most diffusion and coking problems. One of the processes studied was visbreaking, that is, noncatalytic thermal processing. Previous workers had examined the thermal decomposition of an Athabasca asphaltene

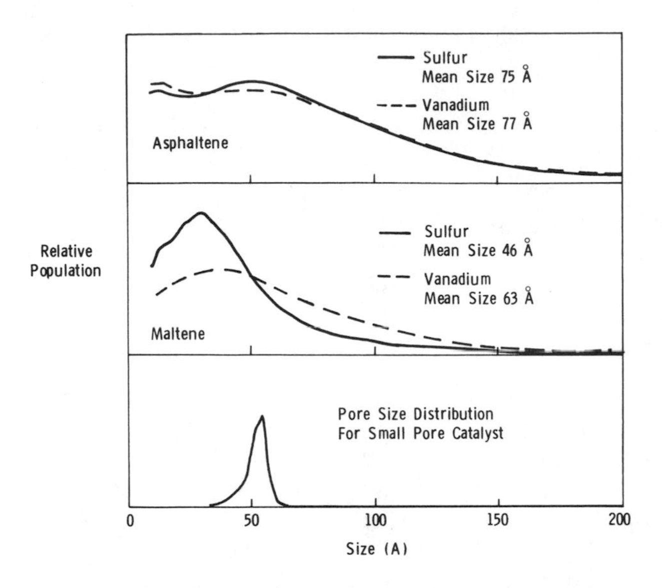

Figure 5. Size distributions for Arabian Light vacuum residuum and small pore catalyst

and found little decomposition below 650°F (*21*). However, above this temperature there was sulfide bond cleavage, evidenced by hydrogen sulfide evolution, and above 750°F thermal cracking of carbon–carbon bonds occurred (*22*). Drushel studied the heat treatment of asphaltenes with GPC and found the molecular weight distribution shifted to lower molecular weights with increasing processing severity (*23*). Figure 6 shows the molecular

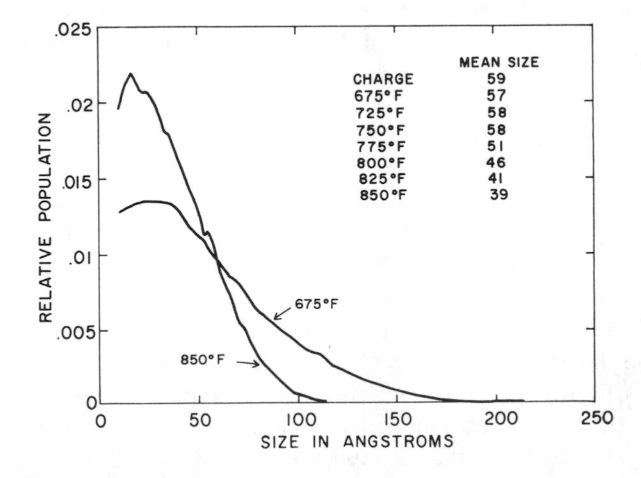

Figure 6. Size distribution of visbroken Arabian Light vacuum asphaltenes

size distributions for Arabian Light vacuum asphaltenes obtained from a residuum that had been visbroken at a series of temperatures from 675°F to 850°F. No significant change in the overall mean size occurs until about 775°F; however, above this temperature a dramatic size decrease is observed. In addition, the wt % asphaltenes value, which is about 13% at temperatures below 750°F, increases to 19% at 850°F, indicating coke formation. Since sulfide bond cleavage would be expected to occur below this temperature, thermal cracking of carbon–carbon bonds and coke formation appear to be more important for the decomposition and size reduction of this particular asphaltene.

Hydroprocessing effects on the molecular size distributions also were monitored. A previous GPC study reported that hydroprocessing caused a pronounced change in the asphaltene molecular weight distribution (23). In this work, an Arabian Light vacuum residuum was examined, and Figure 7 shows the size data for asphaltenes from the raw stock and from products hydroprocessed over a small pore catalyst at three different temperatures. The catalyst selectively processes smaller molecules, which become pentane soluble (the wt % asphaltenes is reduced from 5.4% at 675°F to 2.6% at 775°F) and no longer contribute to the asphaltenes. At the lower temperatures (675°F and 725°F), there is a concentration of the larger molecules in the remaining asphaltenes, with an increase in mean size. Above 750°F, thermal cracking occurs and the mean size is reduced to 46 Å, which is not much smaller than the 51 Å size observed for the straight thermal cracking of this same residuum (*see* 775°F size in Figure 6). Therefore, the predominant processes for the larger asphaltene molecules are external surface reactions (which are an ineffective use of the overall catalyst surface area) at lower temperatures and

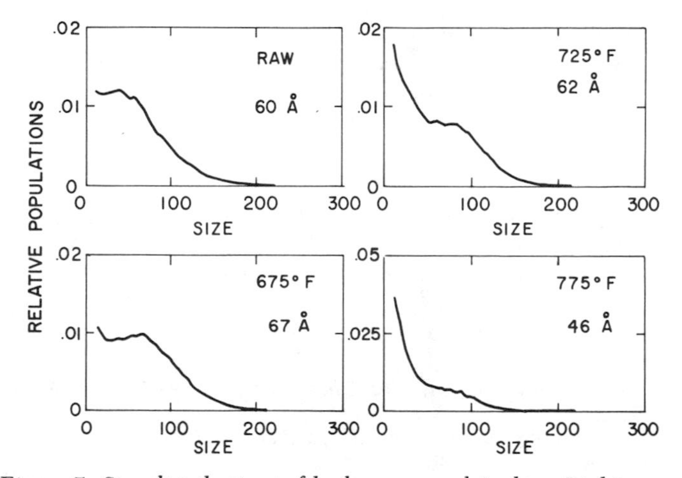

Figure 7. Size distributions of hydroprocessed Arabian Light vacuum asphaltenes

thermal cracking at higher temperatures. Thermal cracking reduces the molecular sizes so the larger molecules can enter the pores and be hydroprocessed, but results in coking, which causes catalyst aging. For this reason the larger asphaltene molecules cannot be processed efficiently over a small pore catalyst.

Comparison of the relative sizes for the nickel- and vanadium-containing molecules in the charge stocks shows some interesting variations. For the Kuwait and Prudhoe Bay residua, the mean size for the nickel compounds is greater than the mean size for the vanadium sizes, while for the Arabian Light vacuum residuum the reverse is true. The mean sizes for the nickel and vanadium compounds are the same for the Lagomedio, Aramco, and Wilmington residua. A wide variety of metal-containing molecules occur in petroleum residua—mainly macrocyclic compounds (both porphyrin and nonporphyrin) containing sulfur, nitrogen, and oxygen atoms bound to the metal (*24*). Since nickel and vanadium have different preferences for specific ligands (N vs. O vs. S) (*25*), it is possible that nickel and vanadium are contained in molecules with different stereochemistries and sizes. The six residua studied here come from several geographic locations and have been subject to different formation and maturation processes, which may have affected their geochemistries (*26*). Therefore, the types and relative sizes of nickel- and vanadium-containing molecules should not be expected to be the same for all six stocks.

Conclusion

Size characterization measurements have provided useful information on the importance of the hydroprocessing catalyst pore size distribution and on the effects of visbreaking and hydroprocessing on the residua molecular size distributions. It is apparent that asphaltenes and maltenes are not unique entities, but instead have considerable overlap in their size distributions. A complete study of the effects of processing conditions would require consideration of all components of a residuum.

Acknowledgment

The authors wish to thank R. B. Callen, S. B. Jaffe and the members of the Hydroprocessing Group, including P. J. Angevine, F. Banta, R. H. Heck, D. Milstein, and T. R. Stein for many helpful discussions.

Literature Cited

1. Yen, T. F. *Am. Chem. Soc., Div. Pet. Chem., Prepr.* (New York, Aug.–Sept., 1972) *17*(4), F102.
2. Speight, J. G. *Appl. Spectrosc. Rev.* **1972,** *5*, 211.
3. Snyder, L. R. *Anal. Chem.* **1969,** *41*, 1223.
4. Altgelt, K. H.; Hirsch, E. *Am. Chem. Soc., Div. Pet. Chem., Prepr.* (Houston, Feb., 1970) *15*(2), A262.

5. Albaugh, E. W.; Query, R. C. *Anal. Chem.* **1976,** *48,* 1579.
6. Gioumatizidou, P.; Oelert, H. H. Erdoel Kohle, Erdgas, Petrochem. Brennst.—Chem. **1977,** *30,* 229.
7. Hendrickson, J. G.; Moore, J. C. *J. Polym. Sci., Part A-4* **1966,** *4,* 167.
8. Simha, R. *J. Appl. Phys.* **1942,** *13,* 147.
9. Reerink, H.; Lijzenga, J. *Anal. Chem.* **1975,** *47,* 2160.
10. Grubisic, Z.; Rempp, P.; Benoit, H. *J. Polym. Sci., Polym. Lett. Ed.* **1967,** *5,* 753.
11. Tung, L. H. *Sep. Sci.* **1970,** *5,* 339.
12. Bergmann, J. G.; Duffy, L. J.; Stevenson, R. B. *Anal. Chem.* **1971,** *43,* 131.
13. Asche, W.; Oelert, H. H. *J. Chromatogr.* **1975,** *106,* 490.
14. Moschopedis, S. E.; Fryer, J. F.; Speight, J. G. *Fuel* **1976,** *55,* 227.
15. Reerink, H. *Ind. Eng. Chem., Prod. Res. Dev.* **1973,** *12,* 82.
16. Drushel, H. V. *Am. Chem. Soc., Div. Pet. Chem., Prepr.* (New York, Aug.–Sept., 1972) *17*(4), F92.
17. Richardson, R. L.; Alley, S. K. *Am. Chem. Soc., Div. Pet. Chem., Prepr.* (Philadelphia, April, 1975) *20*(2), 554.
18. Simpson, C. A. Mobil Research and Development Corp., personal communication.
19. Hastings, K. E.; James, L. C.; Mounce, W. R. *Oil Gas J.* **1975,** June 30, 122.
20. Hardin, A. H.; Packwood, R. H.; Ternan, M. *Am. Chem. Soc., Div. Pet. Chem., Prepr.* (Miami Beach, Sept., 1978) *23*(4), 1470.
21. Moschopedis, S. E.; Parkash, S.; Speight, J. G. *Fuel* **1978,** *57,* 431.
22. Ignasiak, T. M.; Strausz, O. P. *Fuel* **1978,** *57,* 617.
23. Drushel, H. V. *Am. Chem. Soc., Div. Pet. Chem., Prepr.* (New York, April, 1976) *21*(1), 146.
24. Yen, T. F. "The Role of Trace Metals in Petroleum"; Yen, T. F., Ed.; Ann Arbor Science: **1975,** 1.
25. Huheey, J. E. "Inorganic Chemistry, Principles of Structure and Reactivity"; Harper & Row: **1972,** 226.
26. Yen, T. F.; Silverman, S. R. *Am. Chem. Soc., Div. Pet. Chem., Prepr.* (New York, Sept., 1969) *14*(3), E32.

RECEIVED June 23, 1980.

9

Composition of H-Coal Asphaltenes and Preasphaltenes: Acid–Base Separations and Oxidative Degradation

PETER A. S. SMITH, JAMES C. ROMINE, SHANG-SHING P. CHOU, and RICHARD P. SCHRODER

Department of Chemistry, University of Michigan, Ann Arbor, MI 48109

Vacuum-still bottoms from the H-coal liquefaction process were separated into acid, neutral, and basic fractions by precipitation with acids or by extraction with bases. About one-third of the preasphaltene and one-sixth of the asphaltene fraction were precipitated by acids; equivalent weights of the bases were in the range 1200–1800 for preasphaltenes and 600–800 for asphaltenes. The acidic components were obtained either by extraction with aqueous sodium hydroxide or by extraction with benzyltrimethylammonium hydroxide in methanol. About one-fifth of the asphaltene and one-fourth of the presasphaltene fractions were obtained as acids, and up to 10% as amphoteric substances. Nitrogen and sulfur were present in all fractions found. Deno axidation (CF_3CO_2H, H_2O_2, H_2SO_4) gave dicarboxylic acids from malonic to adipic in addition to mono acids.

The nondistillable residue from the H-coal liquefaction process, the vacuum-still bottoms, is a complex mixture of uncertain structure that contains substantial amounts of oxygen, nitrogen, and sulfur in organic form. Schweighardt et al. have shown that some components are sufficiently basic to be precipitated from toluene solution by dry hydrogen chloride (*1*). Nicksic and Jeffries-Harris (*2*) have precipitated petroleum asphaltene bases with perchloric and other acids in glacial acetic acid medium. However, conventional separation of bases by extraction into aqueous acid has not been satisfactorily accomplished because the salts remain more lipophilic than hydrophilic. The precipitated hydrochloride salts are inclined to be hygroscopic and gummy and are not easily handled or purified, a circumstance that

0065-2393/81/0195-0155$05.00/0

makes a material balance subject to large uncertainties. There is also evidence that dry hydrogen chloride may react in ways other than simple salt formation, for the fractions obtained by its use appear to contain some covalently bound chlorine. Furthermore, dry hydrogen chloride is capable of binding bases that would be too weak to form salts in the presence of water.

This work was begun to advance our understanding of the chemical nature of H-coal vacuum bottoms by exploring their interaction with salt-forming reagents. We hoped to achieve improved separations, to provide a basis for chemical fractionation, and to obtain more readily purified salts that would allow good material balances and equivalent-weight determinations of the acidic and basic components. Most of this work was carried out on fractions of the bottoms characterized by solubility. The solvent-extraction procedures, described elsewhere (3), gave four types of materials defined as follows: oils and resins (pentane-soluble); low molecular weight asphaltenes (soluble in 75% pentane/25% toluene; insoluble in pentane); high molecular weight asphaltenes (soluble in toluene, insoluble in 75% pentane/25% toluene), and preasphaltenes (soluble in tetrahydrofuran, insoluble in toluene). As an adjunct to this work, it was desirable to have a means of examining the structural differences among the resulting fractions that would be more revealing than elemental analysis or the somewhat featureless NMR spectra. Oxidative degradation and subsequent analysis of the degradation products by interfaced GC–MS was therefore used to study the aliphatic substituents of the coal-derived materials fractionated by various acid–base separations.

Experimental

Precipitation of Coal-Derived Bases from Toluene. A solution of coal-derived material in toluene was prepared by stirring 50 g of sample per liter of solvent at reflux temperature for 6–8 h. After cooling, suction filtration through a sintered glass funnel was used to remove the undissolved solids, which were dried in vacuo at 80° C and weighed; the quantity of material dissolved was determined by difference. The filtrate was placed in a stoppered bottle and used as a stock solution for several precipitation experiments. Immediately before the use of the toluene solution, an accurately pipetted volume was evaporated to dryness in vacuo and the remaining solid was weighed to calculate the concentration.

An accurately measured volume of the stock solution was placed in an Erlenmeyer flask to which was added an aqueous acid solution (10:1 toluene/acid). The flask was stoppered, placed in an ice bath, and stirred for 1–2 h. For the precipitation of hydrochloride salts, dry hydrogen chloride gas was bubbled through the toluene solution in place of adding aqueous acid (70% perchloric acid, 50% gluconic acid, 1*M* citric acid, or 100% glycerophosphoric acid). The precipitated salts were collected by suction filtration, and were washed three times with toluene, dried in vacuo at 80° C, and weighed.

The isolation of the salts precipitated by glycerophosphoric acid was handled somewhat differently. Filtration of the toluene/acid mixture yielded a gummy, black solid, which was washed with water to give a fine brown precipitate; this was dried in vacuo and weighed. The water washings were treated with aqueous sodium hydroxide and the resulting precipitate was removed by extraction with toluene. The toluene extract was dried (sodium sulfate) and the solvent was evaporated, leaving the free organic bases.

The filtrate and toluene washings from the salt precipitation were combined in a separatory funnel and the aqueous acid layer was removed. The toluene solution was washed three times with 10% aqueous sodium hydroxide and then dried (sodium sulfate). The toluene was evaporated and the residue was dried in vacuo and weighed.

Equivalent Weights. The equivalent weights of the salts precipitated from toluene solution were determined by titration with sodium hydroxide. Some difficulty was encountered in finding a suitable solvent for the titrations, but two methods were developed for obtaining reproducible titration results.

DIRECT TITRATION (METHOD A). An accurately weighed sample of the salt was placed in a small flask containing methanol. Typically, 10 mL of methanol per 0.1 g of salt was used. The flask was lightly stoppered and heated with stirring on a hot plate until the methanol boiled. After cooling, the mixture was titrated with aqueous sodium hydroxide solution. The titration was monitored using a glass combination pH electrode and meter with the endpoint being determined from the resulting titration curve. Because of the incomplete dissolution of the salts in aqueous methanol, the reaction between the salt and the hydroxide titrant was slow. To assure complete neutralization, small aliquots of titrant were added, allowing the pH to stabilize between additions (usually 2–3 min).

INDIRECT TITRATION (METHOD B). A weighed sample of salt was added to a flask containing methanol and a known excess of aqueous sodium hydroxide solution. The mixture was stirred and warmed on a hot plate for 1 h. After cooling, the amount of excess hydroxide present was determined by titration with standard aqueous hydrochloric acid. The titration was monitored using a pH electrode and meter, and the end point was determined from the resulting titration curve.

Recovery of Coal-Derived Bases from their Salts. The salt was accurately weighed and placed in an Erlenmeyer flask with a volume of standard aqueous sodium hydroxide solution sufficient to neutralize it, as calculated from its titrimetric equivalent weight. The mixture was stirred and heated at 50° C for 4–6 h. The pH of the solution was monitored and additional hydroxide solution added when necessary to keep the mixture basic. When the pH had stabilized and no more hydroxide was required, the mixture was filtered through a sintered glass funnel to collect the liberated bases. This solid was washed three times with water, dried in vacuo at 80° C and weighed.

Preparation of β-Glycerophosphoric Acid. The β-glycerophosphoric acid used in the precipitation experiments was freshly prepared before each use. The acid, which decomposes slowly at room temperature, was made from its disodium salt (Reliable Chemical Co.) by ionic exchange in water using Dowex 50W-X8 cation exchange resin. After exchange, the water was removed in vacuo to give the acid as a viscous liquid.

Precipitation of Coal-Derived Bases from Tetrahydrofuran. Weighed amounts (3.0 g) of the following coal-derived materials were each dissolved in 125 mL of reagent grade tetrahydrofuran; preasphaltenes, high molecular weight asphaltenes, low molecular weight asphaltenes, oils and resins, and unfractionated H-coal vacuum-still bottoms. The preasphaltenes and H-coal vacuum bottoms samples did not completely dissolve. The undissolved solids were removed by suction filtration, dried in vacuo, and weighed to maintain a material balance. Each of these tetrahydrofuran stock solutions was treated with phosphotungstic acid in an identical manner; the precipitations were performed in triplicate.

To a 40.0-mL aliquot of stock tetrahydrofuran solution was added a tetrahydrofuran solution of phosphotungstic acid (60 mL, 6.0 g of acid). This mixture was stirred for 4 h at room temperature and allowed to stand for 6 h. The resulting precipitate was removed by suction filtration, washed with tetrahydrofuran (3 × 25 mL), dried in vacuo, and weighed. The organic bases were then recovered from their phosphotungstate salts by warming the salts in a flask containing 25 mL of 10% aqueous sodium hydroxide solution. Suction filtration isolated the resulting free organic bases, which were washed with water, dried in vacuo at 80° C, and weighed.

The organic material not precipitated from the tetrahydrofuran stock solution was recovered by evaporation of the solvent, addition of 100 mL of water, and suction filtration. This precipitate was washed repeatedly with hot water and then placed in a flask with 25 mL of 10% aqueous sodium hydroxide solution to assure complete removal of phosphotungstic acid from the residue. Suction filtration was used to collect the neutralized residue, which was washed with water, dried in vacuo, and weighed. The filtrate from this treatment contained some dissolved organic material, which was recovered by carefully adjusting the solution pH to 4.5 with dilute hydrochloric acid and then extracting with ethyl acetate. The ethyl acetate extract was dried (sodium sulfate) and the solvent was evaporated, leaving a brown solid, which was weighed.

Extraction of Preasphaltenes into Methanolic Triton-B. A sample of preasphaltenes (42.0 g) was placed in a Soxhlet thimble and extracted with methanol for 48 h. The methanol extract was evaporated to dryness to give 4.4 g (10.6%) of methanol-soluble residue. The undissolved solids were removed from the thimble and dried in vacuo. This methanol-insoluble preasphaltene fraction was used in several subsequent experiments.

A sample of methanol-insoluble preasphaltenes (11.4 g) was placed in an Erlenmeyer flask to which was added 100 mL of methanol and methanolic Triton-B (100 mL, 40% w/w). The mixture was stirred for 15 h at room temperature. The undissolved solids were then removed by suction filtration, dried in vacuo, and weighed (5.6 g, 49%). The filtrate was extracted with petroleum ether (10 × 100 mL), and the solvent was removed from the extract, first by rotary evaporation and then in vacuo at 40° C. The petroleum-extractable residue weighed 0.70 g (6%). The methanol solution containing dissolved preasphaltenes was added to 600 mL of water, and the resulting precipitate was collected by suction filtration, washed with water, dried in vacuo at 80° C and weighed (4.0 g). A 0.2173-g sample of this water/methanol-insoluble, Triton-B salt of preasphaltene acids was titrated in methanol with standard aqueous hydrochloric acid to a pH of 4.75; equivalent weight 561 g/equiv. The resulting free acid was collected by suction filtration, dried in vacuo at 80° C, and weighed (0.1689 g, 78%).

A second fraction of organic acids was recovered from the water/methanol filtrate by adjusting the pH to 4.0 with hydrochloric acid followed by suction filtration. The precipitate was washed with water, dried in vacuo at 80° C and weighed (1.6 g, 14%).

Ion Exchange Separation of Asphaltenes

ISOLATION OF ACIDS. Amberlyst A-26 strongly basic anion exchange resin (50 g, Sigma Chemical Co.) was placed in a glass column (4 × 60 cm). The resin was converted from its chloride to hydroxide form by treatment with 10% aqueous sodium hydroxide solution until the effluent gave no precipitation with nitric acid/silver nitrate solution. The resin was then washed successively with water (2 L), methanol (0.5 L), and toluene (0.5 L). An accurately weighed sample (approximately 1 g) of substrate was dissolved in 50 mL of toluene and the solution was placed on the exchange column. Elution with 1 L of toluene removed the nonacidic materials from the resin. This effluent was evaporated, and the residue was dried in vacuo and weighed. The acidic materials were removed from the column by drying the resin with a stream of nitrogen and then eluting with 1 L of toluene/acetic acid (9:1). By drying the resin before changing solvents the problems associated with heat build-up and resin swelling were avoided. The toluene and acetic acid were removed from the collected effluent by rotary evaporation and the resulting asphaltene-acid fraction was dried in vacuo and weighed.

ISOLATION OF BASES. Amberlyst A-15 strongly acidic cation exchange resin (50 g, Sigma Chemical Co.) was placed in a glass column (4 × 60 cm). The resin was washed successively with methanol (0.5 L) and toluene (0.5 L) to remove any soluble resin components. An accurately weighed sample (approximately 1 g) of substrate was dissolved in 50 mL of toluene and the solution was placed on the exchange column, which was then eluted with 1 L of toluene. Rotary evaporation of this effluent yielded a residue of nonbases, which was dried in vacuo and weighed. The basic materials were recovered from the column by drying the resin in a stream of nitrogen and eluting with 1 L of toluene/triethylamine (9:1) mixture. The toluene and triethylamine were removed and the resulting asphaltene-base fraction was dried in vacuo and weighed.

Oxidative Degradation. An accurately weighed sample (approximately 1.6 g) of the material to be oxidized was placed in a 100-mL round-bottomed flask supported in an oil bath at room temperature. To the flask was added hydrogen peroxide (20 mL, 30% solution) and trifluoroacetic acid (10 mL). The flask was fitted with a condenser and the mixture was stirred with a magnetic stirring bar. Over a period of 30 min, sulfuric acid (10 mL, 98%) was added to the oxidation mixture in three portions (slow addition was necessary to prevent overheating). After the addition was completed, the oil bath was heated to 60° C (approximately 15 min was required). All reactions were run for 18 h. After this length of time, the excess peroxide was destroyed catalytically by adding platinum oxide (20 mg) with continued heating and stirring for 6 h.

The oxidative degradation mixture was placed in a large beaker and cooled in an ice bath. Sodium hydroxide solution (50% w/w) was carefully added until the mixture was basic, and it was then transferred to a 1-L round-bottomed flask, which was then attached to a rotary evaporator. The flask was immersed in a Dry Ice/isopropyl alcohol bath and was spun, causing the contents to freeze in a thin layer on the inside of the flask. The flask was then quickly transferred to a vacuum pump where the contents were lyophilized. After water was completely removed, the flask was detached from the pump and 150 mL of methanol were added. Sulfuric acid was then added dropwise until the contents of the flask were acidic. To this mixture was added boron trifluoride/methanol complex (50 mL, $BF_3 \cdot 2CH_3OH$), and the mixture was then heated to 60° C for 18 h. The inorganic solids (sodium sulfate) were removed by filtration and washed with methylene chloride (3 × 20 mL). The filtrate and washings were placed in a separatory funnel to which was added aqueous sodium bicarbonate solution (200 mL, 10% w/w), which caused the contents of the funnel to separate into two layers. The methylene chloride layer was removed and the aqueous layer was extracted with additional methylene chloride. (3 × 25 mL). The extract was dried (sodium sulfate), and octadecane (10.0 μL) was added as an internal reference for GC analysis. The solution was concentrated by fractional distillation through a 60-cm Vigreux column. The resulting methyl ester mixture was then analyzed by GC–MS. After analysis, the ester mixtures were evaporated to dryness and, in some cases, NMR spectra were taken of the resulting residues.

Analysis of Oxidative Degradation Products. The GC–MS analyses of the oxidative degradation products were conducted on a Finnigan Corporation Model 4201 GC–MS with an INCOS 2000 series data system. The resulting data were stored on magnetic disks and compound identification was assisted by a computer program for

Table I. Authentic Compounds Used in GC–MS Analysis

Library Entry Number	*Compound Name*	*Retention Time (min:s)*	*Relative Response Factor*
1	Octadecane	21:54	1.000
2	Dimethyl malonate	8:39	0.249
3	Dimethyl succinate	10:39	0.365
4	Dimethyl glutarate	12:27	0.377
5	Dimethyl adipate	14:06	0.463
6	Methyl propionate	2:45	0.082
7	Methyl isobutyrate	4:06	0.077
8	Methyl valerate	7:09	0.204
9	Dimethyl methylsuccinate	11:24	0.376
10	Methyl benzoate	12:03	0.418
11	Dimethyl phthalate	17:00	—
12	Dimethyl oxalate	6:03	—
13	Dimethyl isophthalate	17:51	—

matching mass spectra of unknown compounds with those stored in a standard spectra library. All analyses were done under the following conditions:

GC Parameters:

Column: 6′ × ¼″ glass; 10% SP-2100 on 80–100 Supelcoport

Oven Temperature: Initial, 40° C for 2 min; Rate of increase 10°/min; final 220° C

Helium Flow Rate: 15 mL/min

Ionization Voltage: 70 V

Photomultiplier Voltage: 1450 V

MS scan rate: 3 s

Mass Range: 45–450 amu

To quantitatively determine the absolute amounts of certain esters present in the oxidation products, the mass spectrometer responses to these compounds were compared with the response to the internal reference, octadecane. Using authentic samples of the esters of interest, relative mass spectrometric response factors were determined. From these response factors, listed in Table I, the amounts of esters present in the oxidative degradation products were calculated using an INCOS subroutine.

Two–Phase Precipitation with Acids

A qualitative survey of the ability of common acids in concentrated aqueous solution to precipitate readily filterable salts from solution of H-coal asphaltenes in toluene suggested perchloric acid as a particularly promising candidate for further examination. The salt that formed at the interface of the solutions was crisply granular, possibly crystalline, although nearly black, and was easily collected and washed. Experiments with this reagent showed that approximately one-third of an asphaltene sample could be precipitated. Because attempts to dry the salts to constant weight by heating led to explosions in some experiments, there is a small uncertainty in the material balances for perchlorate precipitation. Nevertheless, the characteristics of the perchlorates as obtained are not without interest.

Equivalent weights by titration of the asphaltene perchlorates with base fell in the range 570–850; the most reliable values are 715–740. The free bases obtained by decomposing the salts with aqueous sodium hydroxide amounted to about 85% of the weight of the salts, a value that is consistent with the titration values. If the bases contain only carbon, hydrogen, and nitrogen, they must have approximately fifty carbons per nitrogen (or fewer in proportion to oxygen content). Such a ratio approximates the nitrogen content of the original asphaltene, 2.0%, and therefore implies that a substantial portion of the nitrogen of the asphaltene also remained unprecipitated, and is probably nonbasic.

The asphaltene perchlorates are essentially insoluble in water or toluene, but have significant solubility in other solvents: tetrahydrofuran, >0.5%; acetone, 0.39%; chloroform, 0.28%; methanol, 0.17%; benzene, 0.08%. The possibility that extraction with these or other solvents might bring about fractionation of the components is being investigated.

With the preasphaltene fraction of H-coal vacuum-still bottoms, the results were similar, except that the equivalent weights were 1200–1800.

Other acids used for precipitation of preasphaltene bases from a two-phase system are listed in Table II, together with some information about the salts obtained and the bases recovered from them. Citric and glycerophosphoric acids were chosen for their lipophobic anions, which should ensure complete precipitation or extraction from the hydrocarbon solvents.

The most obvious generalization implied by the data is that several acids precipitate about one-third of the preasphaltene, and that in those cases the nitrogen content of the bases is equal to or greater than that of the original material. Thus, it seems to be a soundly based conclusion that preasphaltenes consist of 30%–35% of components that are distinctly stronger bases than water, and that a substantial part of the nitrogen is bound in structures of lower basicity.

Table II. Survey of Acids for Precipitation of Preasphaltene Bases from Toluene

Acid	Unprecipitated Residue (%)	Material Balance: Salt (%)	Free Base (%)	Total (%)	Equivalent Weight (g/equiv.)
Hydrochloric	56	37	36	92	959
Citric	70	38	33	103	1490
Glycero-phosphoric	78	32	28 (5)[a]	111	1420, 1320

[a]Obtained from the water-soluble salts by basification of the aqueous phase.

Table III. Elemental Analyses of Preasphaltene Bases

Sample	*C (%)*	*H (%)*	*N (%)*	*S (%)*	*C/N*	*C/S*	*N/S*
Original	84.70	5.70	2.04	1.11	35.6	28.6	0.91
Toluene-insoluble	84.47	4.70	2.03	1.20	35.7	26.4	0.77
Toluene-soluble	84.70	5.70	2.04	1.11	35.6	28.6	0.83
Bases from HCl[a]	81.44	5.76	2.33	0.51	30.0	60.0	2.0
Bases from citric	74.09	4.60	1.91	0.54	33.2	51.4	1.5
Bases from glycerophosphoric, water insoluble	86.35	5.95	3.12	0.39	23.7	83.0	3.5
Bases from glycerophosphoric, water soluble	81.30	6.28	4.24	1.76	16.4	17.3	1.0
Not precipitated by HCl[b]	84.90	7.23	1.44	0.33	50.5	96.5	1.9
Not precipitated by citric	86.52	6.94	1.10	0.15	67.4	216.0	3.2
Not precipitated by glycerophosphoric	81.37	7.02	0.82	0.32	85.1	95.3	1.1

[a] 3.25% Cl.
[b] 1.68% Cl.

Glycerophosphoric acid proved to be interesting in that approximately 5% of the preasphaltene bases had water-soluble glycerophosphate salts. It appears that the highly hydrophilic glycerophosphate anion can act as a phase-transfer agent for a fraction of preasphaltene bases that must be distinguished from the other H-coal fractions in some structural way.

In Table III are reported elemental analyses of the original sample (toluene-soluble and toluene-insoluble fractions), of the bases obtained from them, and of the unprecipitated remainder. The preasphaltenes, though defined as tetrahydrofuran-soluble, toluene-insoluble materials, are sufficiently soluble in toluene (approximately 1 g/100 mL) to be used in these precipitation experiments. The elemental analysis data suggest that no significant fractionation has taken place in this dissolution step.

An important feature of these results concerns the sulfur content. A large part of the sulfur accompanies the basic fraction. This fact could reasonably be explained by the presence of components containing the thiazole ring (probably as benzothiazoles). Furthermore, comparison of the sulfur content of the bases and the nonbasic residues with that of the original material shows a net decrease. This may, in fact, be attributable to the removal of water-soluble inorganic sulfur compounds during the precipitation procedure.

From the data in Table III, it is evident that preasphaltene bases fractionated as water-soluble glycerophosphate salts show a remarkable concentration of both nitrogen and sulfur. This material was recovered from the aqueous filtrate in the glycerophosphoric acid experiment by addition of alkali, and therefore, is not amphoteric. Although these bases represent only about 5% of the preasphaltene materials, they illustrate that fractionation is feasible by chemical means, and we are currently exploring the scope of the method.

Table III includes the C/N, C/S, and N/S ratios for the various fractions to make the changes brought about by acid–base fractionation more apparent. The fact that the N/S ratio varies widely makes it clear that although sulfur is associated with nitrogen in some of the basic species, there must be other basic species that contain nitrogen without sulfur.

One-Phase Precipitation: Phosphotungstates

In addition to concerns over the incomplete solubility of some of the H-coal vacuum-still bottom materials in toluene, the precipitation experiments just discussed are subject to problems associated with two-phase reactions in general. Incomplete contact between the organic compounds in toluene solution and the aqueous acids can lead to inconsistent precipitation results. With this in mind, a different approach was developed in which various H-coal vacuum-still bottoms were treated with phosphotungstic acid, which is soluble in tetrahydrofuran. This reagent allows precipitation to be accomplished from an initially one-phase system, and gives rise to crisp, easily collected salts.

Table IV. Precipitation of Coal-Derived Bases with Phosphotungstic Acid

	Free Bases (%)[a]	*Neutral Residue, (%)*	*Acids from NaOH (%)*	*Total*
Asphaltenes	17.7	63.0	20.0	100.7
	17.3	61.0	20.6	99.0
	18.5	59.0	21.6	99.4
Preasphaltenes	34.7	29.5	25.3	89.6
	33.7	31.7	24.8	90.2
	34.8	31.3	23.9	90.0
H-Coal vacuum-still bottoms, unfractionated	3.5	71.3	16.1	91.1
	2.4	72.2	17.7	92.4
	2.7	73.5	17.4	93.6

[a]Percentages are weight compared with the original sample.

The results from these experiments are presented in Table IV for three coal-derived materials: high molecular weight asphaltenes, preasphaltenes, and unfractionated H-coal vacuum-still bottoms. After collection of the phosphotungstate salts, the tetrahydrofuran filtrate was evaporated and treated with aqueous sodium hydroxide to remove excess phosphotungstic acid. Surprisingly, 16%–25% of this nonbasic residue was soluble in the aqueous alkali, a reagent that does not appreciably dissolve the original material. This alkali-soluble fraction is listed as "acids" in Table IV, while the alkali-insoluble portion is given as neutral residues.

The phosphotungstate salts cannot be used for equivalent weight determinations, for their characteristically variable composition leads to inconsistent titration results. However, treatment of the salts with aqueous sodium hydroxide led to recovery of the free bases in reproducible amounts. These results show that one-fifth of the high molecular weight asphaltenes and one-third of the preasphaltenes are precipitated as phosphotungstate salts. Considerably less of the unfractionated H-coal vacuum-still bottoms and none of the low molecular weight asphaltenes and oils and resins were separated in this manner.

The fact that part of the nonbasic fractions of the asphaltene or preasphaltene samples could be dissolved in aqueous sodium hydroxide was surprising, for when untreated samples are extracted with aqueous base, almost nothing is removed. Removal of the nitrogen bases evidently frees the acidic components for reaction with hydroxide, even though the bases are only weakly basic and should not be able to compete with hydroxide ion for the acidic components. Presumably this is a question of wetting, contact, or occlusion.

The percent of alkali-soluble acids varies somewhat, but not greatly, among asphaltenes, preasphaltenes, and total H-coal vacuum bottoms, but the percent of precipitable bases varies markedly. Assuming that phosphotungstic acid scavenges essentially all of the bases, the change from preasphaltenes to asphaltenes involves a cleavage of neutral material from large, basic molecules, thereby producing about half the weight of bases, presumably having about half the molecular weight. However, these must be regarded as tentative conclusions, for our experiments do not cover a representative group of samples. However, they are consistent with the fact that the equivalent weights of other salts of asphaltene bases (e.g., perchlorates) are about half the equivalent weight of preasphaltene base salts.

Extraction with Bases

Husack and Golumbic (*4*) reported the isolation of the phenolic acids from asphaltenes by extraction into Claisen alkali (KOH in water/methanol). In a somewhat similar approach, H-coal preasphaltenes were treated with several methanolic hydroxide solutions. The results, shown in Table V, show that 10%–12% of the preasphaltenes are soluble in methanol alone. Substantially more of these materials dissolve in 1.5*M* sodium or lithium hydroxide solutions. The greatest amounts of preasphaltenes were extracted by 1.5*M* solutions of quaternary ammonium hydroxide bases. Because Triton-B (benzyltrimethylammonium hydroxide) in methanol was capable of dissolving 65% of the preasphaltene materials, it was used in a more elaborate fractionation scheme.

A sample of methanol-insoluble preasphaltenes (89.4% of the starting preasphaltenes) was treated with 1.5*M* methanolic Triton-B; 49% of undissolved material remained. Extracting the filtrate with petroleum ether gave 6% of a nonpolar fraction. Water added to the methanolic phase precipitated insoluble Triton-B salts, which yielded 28% of acids upon acidification. The filtrate from these salts, largely aqueous, contained soluble Triton-B salts, from which a further 14% of acids was obtained. The material balance of the methanol-insoluble preasphaltene was thus 97%. The elemental analyses of these fractions are shown in Table VI.

It can be seen from Table VI that not only is sulfur associated with the acidic components, but its concentration is higher in them than in the

Table V. Separation of Preasphaltenes by Extraction with Basic Methanol

Base	*Percent Undissolved*	*Percent Recovered from Methanol Extract*	*Total*
None	89.4	10.6	100
	82.0	12.0	94
NaOH	65.0	27.0	92
LiOH	57.0	33.0	98
$(C_2H_5)_4N^+OH^-$	54.0	56.0	110
Triton-B	32.0	65.0	97

Table VI. Elemental Analyses of Preasphaltene Fractions Obtained with Triton-B in Methanol

Fraction	*C*	*H*	*N*	*S*	*(O)*[a]	*N/S*
Petroleum ether solubles	86.70	11.00	0.79	0.42	1.09	0.75
Undissolved material	72.18	6.17	4.03	0.72	16.90	2.24
Free acid, water insoluble	77.97	5.41	1.91	1.00	13.71	0.76
Free acid, water soluble	65.96	7.03	3.91	0.85	22.35	1.84

[a]Oxygen calculated by difference.

nonacids. The wide variations in the N/S ratios and in the oxygen content show that the acidic components are subject to fractionation by differential solubility of salts.

The fractions separated using the Triton-B scheme were analyzed by proton NMR. With the exception of the petroleum ether solubles, this acid–base separation did little to fractionate the preasphaltenes based on proton NMR character. The results are given in Table VII.

Comparison with Ion Exchange Separation

Other workers (*5*, *6*) have reported the acid–base fractionation of asphaltene materials by ion exchange chromatography. To compare the results reported here with those obtained in this chromatographic method, H-coal high molecular weight asphaltenes were fractionated by ion exchange, using Amberlyst A-26 anion exchange resin and Amberlyst A-15 cation exchange resin. The results are given in Table VIII. The experiment was conducted in two ways, one removing the bases first and the other removing the acids first.

Table VII. Proton NMR Data from the Acid–Base Separation Fractions

	Absorptions (ppm)		
Material	*Aliphatic*	*Aromatic*	*Aliphatic to Aromatic Ratio*
Preasphaltene 42 (4), methanol soluble	0.7–3.0	7.5–8.2	2.1
Methanol insoluble	0.7–4.4	6.0–9.0	2.5
Triton-B scheme, undissolved material	0.7–4.6	6.0–9.0	2.6
Petroleum ether solubles	0.7–2.7	7.0–7.4	16.6
Free acids, water insoluble	0.7–4.8	6.0–9.0	2.8
Free acids, water soluble	0.7–4.4	6.0–9.0	2.7

Table VIII. Fractionation of High Molecular Weight Asphaltenes by Ion Exchange Chromatography

	Bases (%)	*Neutrals (%)*	*Acids (%)*	*Total*
Bases Removed First	23.7	62.4	13.0	99.1
Acids Removed First	14.8	63.5	23.6	101.9
Difference	8.9	1.1	10.6	

The differences reflected in the two methods indicate the presence of amphoteric asphaltene compounds, an observation that is at odds with that reported by Sternberg et al. (*7*) for asphaltenes from the Synthoil process. The bases isolated by this ion exchange method correspond closely in amount to those isolated by the phosphotungstic acid scheme.

In a procedure similar to the one just described, an attempt was made to fractionate preasphaltenes. Because of the low solubility of preasphaltenes in toluene, tetrahydrofuran was chosen as the solvent for this ion exchange chromatography. The results where the bases were removed first show 40.2% of the material to be basic and 89.7% to be neutral and acidic. The excessive material recovery (129.9%) was determined to be the result of a partial dissolution of the exchange resin in tetrahydrofuran. Similar limitations of the ion exchange method, discussed by others, emphasize the usefulness of the precipitation methods already discussed (*8, 9*).

Oxidative Degradation

Further analysis of the H-coal materials fractionated by acid–base chemistry was performed using an oxidative degradation method developed by Deno and coworkers (*10*). In this method, a mixture of trifluoroacetic acid/hydrogen peroxide/sulfuric acid can be used to selectively destroy the aromatic ring in a sample while leaving the aliphatic substituents unoxidized. For example, the result of the oxidation of ethylbenzene by this method would be propionic acid while diphenylmethane would react to give malonic acid. This oxidative degradation allows one to investigate the aliphatic components of coal-derived materials, which in this case were identified by GC–MS as methyl ester derivatives of the degradation products.

The degradation products from the oxidation of a number of H-coal samples were qualitatively quite similar. The major compounds identified as methyl esters included: propionate, butyrate, oxalate, malonate, methylmalonate, maleate, succinate, methylsuccinate, citraconate, glutarate, dimethylsuccinate, methylglutarate, adipate, methylethylmaleate, methyladipate, methoxysuccinate, benzoate, naphthoate, phthalate, and terephthalate.

To characterize the H-coal samples separated by acid–base chemistry, a quantitative analysis of several oxidative degradation products proved to be useful. Chosen for this purpose were the methyl esters of malonic, succinic,

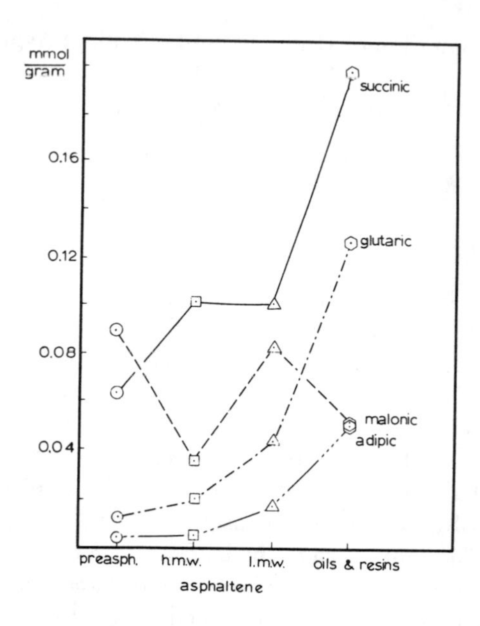

Figure 1. Degradation products from solubility-characterized H-coal fractions: h.m.w. = high molecular weight; l.m.w. = low molecular weight

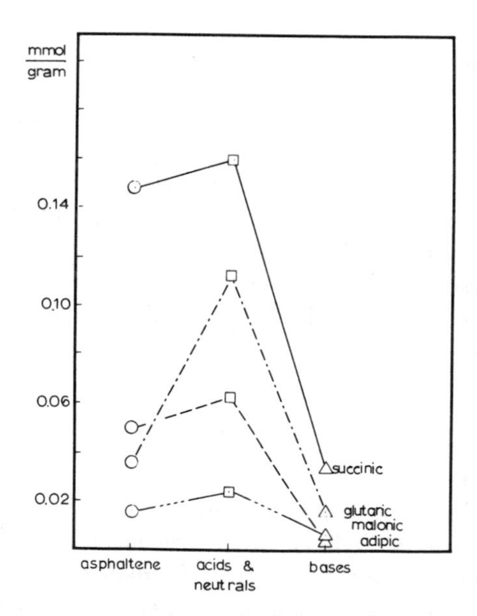

Figure 2. Degradation products from asphaltene fractions separated by phosphotungstic acid

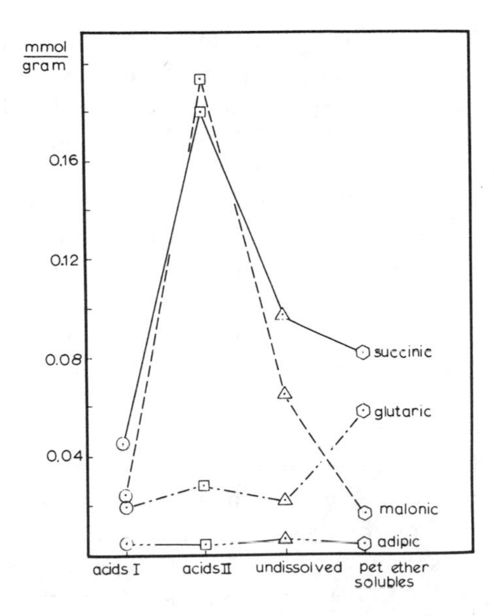

Figure 3. Degradation products from preasphaltene fractions separated by Triton-B: acids I, acidic fraction from water-soluble salts; acids II, acidic fraction from water-insoluble salts

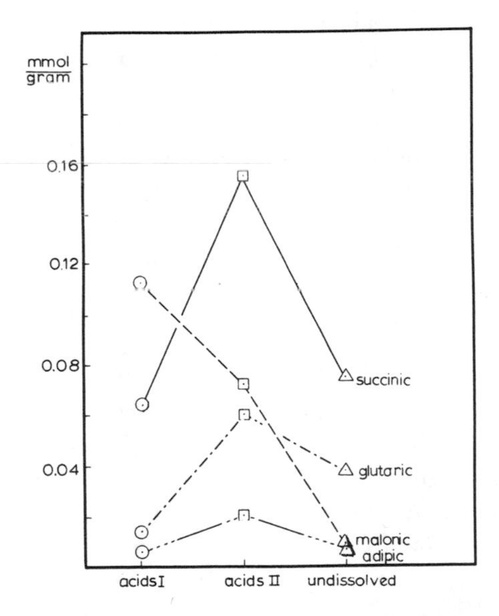

Figure 4. Degradation products from asphaltene fractions separated by Triton-B: acids I, acidic fraction from water-soluble salts; acids II, acidic fraction from water-insoluble salts

glutaric, and adipic acids. These products represent methylene chains of between one and four carbons present in the sample oxidized. These types of aliphatic structures are important both because of their relative abundance and because of their possible role in linking the aromatic subunits present in the coal-derived materials.

The results from GC–MS analysis of dimethyl malonate, succinate, glutarate, and adipate in degradation products of several series of H-coal materials are given in Figures 1–4. The graphs in these figures show the amounts of diester products (in mmol/g of sample oxidized) on the abscissa for a series of samples given on the ordinate. The lines connecting these points allow one to see the relationship between the amounts of these key degradation products and the scheme used to separate the series of coal-derived samples.

From Figure 1, one can see an increase in the amounts of the diester degradation products with decrease in apparent polarity of the H-coal fractions as determined by their solubility characterization. Figure 2, which results from asphaltenes fractionated by phosphotungstic acid, indicates that a greater amount of the products are produced from the oxidation of the acid–neutral material. Figures 2 and 4 both indicate that larger amounts of diester are found in the degradation products of the free acids of preasphaltenes and asphaltenes as isolated by the Triton-B scheme.

One can imagine two plausible explanations for the observation that acidic fractions from II-coal consistently give greater amounts of diester products: either the acidic fractions (which are phenolic) are more reactive and, thus, oxidize more completely than do the basic fractions, or the acidic components actually contain a greater number of methylene chains than do the bases. Either condition would be reflected in the amounts of diesters produced by oxidative degradation. This type of oxidative analysis provides both a means for characterizing the H-coal fractions and information that can be used to refine hypotheses about the aliphatic structures present in these materials.

Acknowledgment

This research was carried out under contract no. EX-76-S-01-2550 with the United States Energy Research and Development Administration and a preliminary research grant from the University of Michigan, Division of Research Development Administration. A large part of it is taken from the doctoral dissertation of James C. Romine, University of Michigan, 1980.

Literature Cited

1. Schweighardt, F. K.; Retcofsky, H. L.; Raymond, R. *Am. Chem. Soc., Div. of Fuel Chem., Prepr.* (San Francisco, Aug.–Sept., 1976) *21*(7), 27.
2. Nicksic, S. W.; Jeffries-Harris, M. J. *J. Inst. Pet.* **1968,** *54,* 017.
3. Addington, D. V., Ph.D. Thesis, Univ. of Michigan, 1978.

4. Husack, R.; Golumbic, C. *J. Am. Chem. Soc.* **1951,** *73,* 1567.
5. Boduszynski, M.; Chadha, B. R.; Szkuta-Pochopien, T. *Fuel* **1977,** *56,* 432.
6. Farcasiu, M. *Fuel* **1977,** *56,* 9.
7. Sternberg, H. W.; Raymond, R.; Schweighardt, F. K. *Science* **1975,** *188,* 49.
8. Ruberto, R. G.; Jewell, D. M.; Cronauer, D. C. *Fuel* **1978,** *57,* 575.
9. Farcasiu, M. *Fuel* **1978,** *57,* 576.
10. Deno, N. C.; Greigger, B. A.; Stroud, S. G. *Fuel* **1978,** *57,* 455.

RECEIVED June 23, 1980.

10

Molecular Interactions Involving Coal-Derived Asphaltenes

KRISHNA C. TEWARI[1] and NORMAN C. LI

Department of Chemistry, Duquesne University, Pittsburgh, PA 15219

Interaction of quinoline (Qu) with coal-derived asphaltenes (A), acid/neutral (AA) and base (BA) components of A, silylated asphaltenes A(TMS), and pentane-soluble heavy oil (HO) fractions, obtained from same feed coal, in solvent benzene has been studied by calorimetric method. The linear variation of molar enthalpy (for a 1:1 complexation) with the phenolic oxygen content of the fractions has been attributed to the dominance of hydrogen bonding effects involving phenolic OH over other types of molecular interactions. In Qu-A(TMS) systems, the degree of complexation largely depends on some entropy effects. For interaction between HO and asphaltenes in benzene, both viscosity and molar enthalpy change in the order BA > A > AA. These correlate with the NMR downfield shift of o-*phenylphenol–OH signal as a function of asphaltene (A, AA, BA) concentration and suggest hydrogen bonding involving largely phenolic OH as a mechanism by which asphaltene–HO interactions are achieved. Other measurements with amine mixtures indicate that hydrogen bonding involving phenolic–OH and nitrogen bases gives rise to proton-transfer complexes, which partially accounts for the high viscosity and non-Newtonian flow of the coal liquids.*

The high viscosity at ambient temperature of coal liquids derived from hydrogenation processes has been related to the asphaltene (toluene-soluble, pentane-insoluble) and preasphaltene (toluene-insoluble, pyridine-soluble) fractions (*1–5*). Although the effect of preasphaltene concentration on the viscosity of coal liquids is dramatic, the increase caused by asphaltene materials has been attributed to hydrogen-bonding (*6*) and acid–base salt

[1]Current address: Air Products and Chemicals, Inc., Box 538, Allentown, PA 18105.

0065-2393/81/0915-0173$05.00/0

formation (*2, 4*) interactions. We report here calorimetric and NMR results on the hydrogen-bonding interaction involved in coal-derived subfractions and its effect on the viscosity of coal liquids.

Experimental

Centrifuged liquid product (CLP) samples, FB53 batch 1 and 59, and FB57 batch 42 were made in the Process Development Unit at Pittsburgh Energy Technology Center, under different process conditions (*7*) from the same feed coal, Kentucky hvAb, from Homestead Mine. SRC II process solvent (referred to as SRC II) was prepared by Pittsburgh and Midway Coal Company, from Kentucky #9 and #14 blend bituminous coal. The isolation of toluene-insoluble (TI), asphaltene (A), and heavy oil (HO) from CLP was accomplished by solvent fractionation based upon solubility in toluene and pentane (*7*). The A fraction was further separated into acid/neutral (AA) and base (BA) components by bubbling hydrogen chloride gas through a stirred toluene solution (*8*). The same method was used to separate SRC II into acid/neutral (SRC II–acid) and base fractions. A 2:1 (v/v mixture of hexamethyldisilazane and *N*-trimethylsilyldiethylamine was used for the hydroxyl silylation of asphaltene (A, AA, BA) and HO samples (*9*). NMR integration and elemental analysis data were used to calculate the phenolic oxygen content of these fractions. The Bolles and Drago (*10*) calorimetric method was used to determine simultaneously the molar enthalpy, ΔH°, and equilibrium constant, K, for a 1:1 complexation in a donor–acceptor type of reaction $A + B \rightleftharpoons C$. The applicability of the calorimetric method and reliability of the assumed 1:1 complex in systems involving coal-liquid fractions have been described previously (*7, 8, 9*).

Results and Discussion

The viscosities (*11*) at 355 K of the three CLP samples, FB53-1, FB57-42, and FB53-59 derived from the same feed coal, were 25.1, approximately 128, and greater than 700 saybolt seconds, respectively. The results of solvent fractionation, ultimate analyses, and molecular weight determinations (VPO) of the fractions are given in Table I (*7, 8*). The NMR proton distribution in carbon disulfide solutions and calculated structural parameters of the fractions, using Brown and Ladner (*12*) equations, are shown in Table II (*7*). The results indicate that the lower viscosity liquid product contains a lower weight percent of TI, A and a higher weight percent of BA in A. The A, AA, BA and HO fractions isolated from low viscosity liquid contain lower oxygen contents than do fractions from high viscosity liquid product. Furthermore, in agreement with the trend of C/H ratio (*see* Table I), the structural parameters such as aromaticity, f_a, and aromatic/benzylic hydrogen ratio are the same for the three HO fractions and decrease in the order of FB53-1 > FB57-42 > FB53-59 for the A fractions.

Quinoline Interaction With A, AA, BA and HO Fractions

The phenolic oxygen contents in A, AA, BA and HO fractions and the thermodynamic parameters of their interactions with quinoline (Qu) in solvent benzene are summarized in Table III (*13*). For a given system, Qu + A

Table I. Weight Distribution and Ultimate Analysis of Coal-Liquid Fractions[a]

Source	Fraction	Weight Percent Ash Free	Molecular Weight	Atomic Ratio C/H	O/C	N/C	S/C
FB53-1	TI	5.6	—	1.45	0.055	0.023	0.010
	A	19.0	680	1.25	0.026	0.019	0.003
	HO	75.4	240	0.84	0.014	0.008	0.002
	A (TMS)	—	770				
FB57-42	TI	9.3	—	1.33	0.082	0.028	0.033
	A	28.4	530	1.12	0.036	0.020	0.004
	AA	13.3	430	1.06	0.041	0.012	0.003
	BA (HCl-free)	15.1	680	1.18	0.030	0.030	0.005
	HO	62.3	260	0.85	0.025	0.011	0.003
	A (TMS)	—	660				
FB53-59	TI	10.4	—	1.36	0.120	0.024	0.051
	A	33.3	740	1.10	0.038	0.020	0.004
	AA	17.7	620	0.97	0.049	0.011	0.004
	BA (HCl-free)	15.6	950	1.08	0.037	0.028	0.003
	HO	56.3	290	0.84	0.028	0.010	0.003
	A (TMS)	—	830				
SRCII		100	190	0.86	0.027	0.013	0.002
	SRC II-Acid	90.7	210	0.86	0.027	0.008	0.001

[a]Ref. (7, 8).

Table II. Proton Distribution and Structural Parameters of Coal-Liquid Fractions (7)

Source	Fraction	Area Percent NMR Spectra: Aromatic (H_a)	Benzylic[a] (H_α)	Aliphatic[a] (H_o)	f_a	σ	H_{au}/C_a
FB53-1	A	38.8	30.3	30.9	0.75	0.32	0.61
	HO	23.7	31.1	45.2	0.55	0.41	0.87
FB57-42	A	36.5	33.5	30.0	0.72	0.36	0.72
	HO	25.8	32.9	41.3	0.56	0.42	0.93
FB53-59	A	32.6	34.5	32.9	0.70	0.40	0.70
	AA	25.6	22.9	51.6	0.62	0.39	0.70
	BA	32.9	32.4	34.7	0.69	0.38	0.71
	HO	23.8	31.9	44.3	0.54	0.43	0.93

[a]Separation point between H_α and H_o chosen at 1.94 ppm from TMS.

Fuel

and Qu + HO, the values of the equilibrium constant, K, are the same within experimental error while the molar enthalpies of interaction, $\Delta H°$, increase markedly with the increase in oxygen content of the coal-liquid fraction (in the order FB53-1 < FB57-42 < FB53-59). Coal-liquid fractions are complex mixtures of substituted heterocyclic aromatics; therefore, the observed values of K and $\Delta H°$ correspond to the total interaction involving hydrogen-bonding and other types of rapidly reversible intermolecular interactions. Since C/H ratio, aromaticity (f_a), and other structural parameters are the same for the three HO fractions (*see* Table II), the π-bonding contributions to the observed $\Delta H°$ values of Qu + HO systems, to a large extent, would be the same. Since the C/H ratio and f_a decrease in the order of FB53-1 > FB57-42 > FB53-59 for the A fraction, the computed K values for the interaction of Qu with A, AA, and BA fractions are almost the same, and the C/H ratio decreases in the order BA > A > AA, the observed rectilinear variation of $\Delta H°$ with the phenolic oxygen content of coal-liquid fractions, Figure 1 (*13*), could be attributed to the dominance of hydrogen-bonding effects, involving phenolic OH, over other types of molecular interactions in solution.

In the absence of phenolic OH, the calculated K values for the interaction of Qu with silylated asphaltenes [A(TMS), Table III] are not the same and are quite small compared with those observed for the Qu + A system. In the Qu + A(TMS) systems the results indicate that although the strength of interaction ($\Delta H°$) is appreciable, the degree of complexation largely depends upon some unusual entropy effect. It is difficult to speculate on the factors influencing entropy; however, the observed $\Delta H°$ values increase with decreas-

Table III. Hydroxyl Distribution in Coal-Derived Liquid Fractions and Thermodynamic Parameters of Their Interaction with Quinoline in Solvent Benzene at 298 ± 0.5 K[a]

			Thermodynamic Constants of Quinoline Interaction		
Source	*Fraction*	*Phenolic Oxygen (g/mol)*	K^{-1} *(mol dm^{-3})*	$-\Delta H°$ *(kj mol^{-1})*	$-\Delta S°$ *(j mol^{-1} K^{-1})*
FB53-1	A	12.1	0.0515	14.98 ± 0.13	25.5
	HO	3.0	0.0323	4.23 ± 0.04	−14.2
	A (TMS)	—	0.6309	13.31 ± 0.25	40.8
FB57-42	A	14.3	0.0532	16.92 ± 0.21	32.2
	AA	12.5	0.0549	14.74 ± 0.04	25.3
	BA	9.4	0.0543	11.77 ± 0.04	15.2
	HO	6.1	0.0328	7.49 ± 0.12	−3.3
	A (TMS)	—	0.2354	14.14 ± 0.22	35.5
FB53-59	A	22.3	0.0585	26.02 ± 0.13	63.6
	HO	7.6	0.0352	8.28 ± 0.08	0.0
	A (TMS)	—	0.4553	10.84 ± 0.35	29.8

[a]Ref. (*13*).

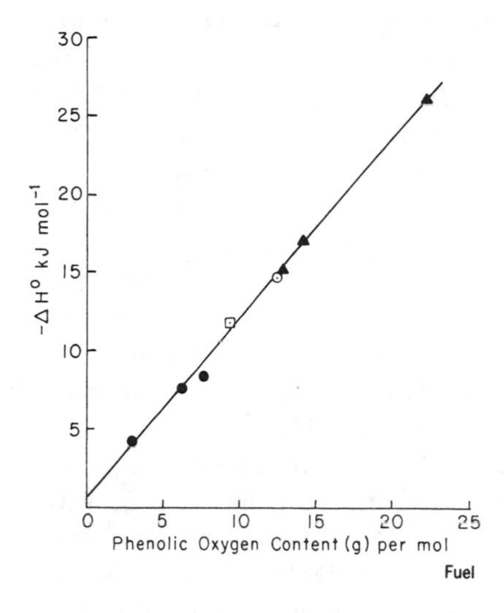

Figure 1. Dependence of ΔH^o on the phenolic oxygen content of the coal fraction: (▲) A, (○) AA, (□) BA, (●) HO (13)

ing molecular weight (*see* Table I) and, therefore, decreasing size of the A(TMS) "molecule." The decrease in entropy of the Qu-A(TMS) complex with the increase in aromaticity, f_a, of the A fraction (Table II) indicate the possible restriction associated with the Qu molecule joining the larger polynuclear condensed aromatic framework of the A(TMS) "molecule."

Interaction Between Asphaltene and Heavy Oil Fractions

The relative viscosity (η_γ) changes for benzene solutions of HO (0.305*M*) at 293 K, with added AA and BA isolated from FB57-42, are shown in Figure 2 (Curves 2 and 5). The results indicate that at a given concentration (above 0.035*M*) and temperature, BA has a larger effect on viscosity than does AA.

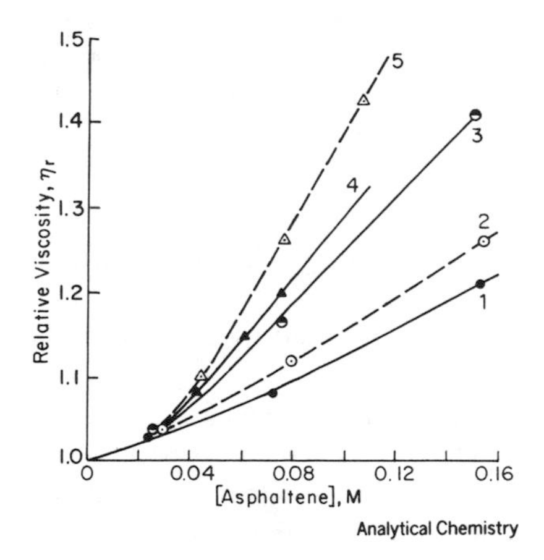

Figure 2. Relative viscosity change with asphaltene concentration in C_6H_6 and in 0.305M solution of HO in C_6H_6 at 293 K. Sample FB57-42: (●) AA in C_6H_6, (○) AA in HO + C_6H_6, (▲) BA in C_6H_6, (△) BA in HO + C_6H_6; Sample FB53-59: (◒) AA in C_6H_6 (9).

Table IV. Summary of Thermodynamic Constants[a] at 298 ± 0.5 K (9)

Source	*System*	K ($dm^3\ mol^{-1}$)	$-\Delta H°$ ($kj\ mol^{-1}$)
FB53-59	HO + A in C_6H_6	9.3	19.12 ± 0.57
	HO + AA in C_6H_6	9.1	15.02 ± 0.48
	HO + BA in C_6H_6	9.4	25.90 ± 0.78

[a]Uncertainties in $\Delta H°$ values are standard deviations. Error in K is about 10%.

Analytical Chemistry

Since coal-derived asphaltenes are known to associate even in dilute solution and, more significantly, in nonpolar solvents (*14, 15*), the effect of molecular size on the observed viscosity can be seen qualitatively from the variation of η_γ with concentration of added asphaltene fraction in pure benzene (Figure 2, Curves 1, 3 and 4). At a given temperature and concentration, η_γ varies linearly with the molecular weight of the added fraction, and for a given fraction (AA or BA) the value of η_γ in benzene is smaller than that for the same fraction in benzene containing HO. Furthermore, in the presence of HO and at a given concentration of asphaltene, the increase in η_γ for BA is larger than that for AA. Since solvent benzene is less polar than benzene containing HO, it is assumed that in addition to the molecular weight of the added asphaltene fraction, part of the effect on viscosity is attributable to functional groups such as phenolic or alcoholic hydroxyl as well as acidic NH groups, which serve as hydrogen donors in intermolecular association. The contribution involving pyrrol type of imino groups as hydrogen donors is negligible since the pK_a of phenol and pyrrol in aqueous solution at 293 K are 9.89 and 15, respectively.

The thermodynamic constants for the interaction of A, AA, and BA with HO in benzene are summarized in Table IV. The computed K values, within experimental error, are the same while $\Delta H°$ increases markedly with the increase in molecular weight and nitrogen content and the decrease in oxygen content of the asphaltene fraction (in the order AA $<$ A $<$ BA). The computed K values also show a direct correlation with the viscosity results shown in Figure 2. Since the C/H ratio and NMR structural parameters of both A and BA (Table II) are the same, the π-bonding contribution to the observed $\Delta H°$ values of HO interaction with A and with BA would be largely the same. The observed increase of $\Delta H°$ value in the order AA $<$ A $<$ BA, therefore, is attributable to the degree of hydrogen-bonding basicity of these fractions. In addition, the hydrogen-bonding interactions involving phenolic hydroxyl protons can be seen from the NMR downfield chemical shift of the phenolic OH signal of *o*-phenylphenol (OPP). OPP is a model used to study phenolic OH in HO as a function of added aslphaltene (A, AA, BA) in solvent CS_2 (Figure 3). The observed shift of the OPP–OH signal at a given asphaltene (A, AA, BA) concentration is fairly large, downfield in the order BA $>$ A $>$ AA, and correlates well with the viscosity and calorimetric results reported above. Furthermore, addition of silylated asphaltene, A(TMS), into 0.2*M* OPP in CS_2

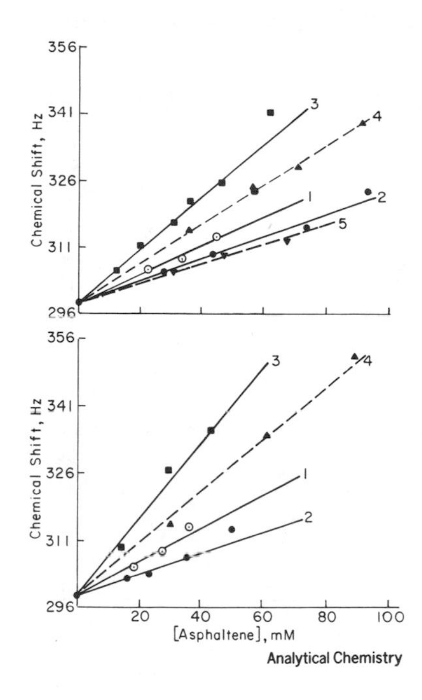

Figure 3. Chemical shift changes at 60 MHz of the proton–OH resonance of o-phenylphenol (0.2M) as a function of asphaltene concentration in CS_2 solution; added asphaltene: (○) A, (●) AA, (■) BA, (▲) A(TMS), (▼) AA(TMS) (9).

(Figure 3) moves the OPP–OH signal considerably downfield compared with that observed for A addition. This is expected because the removal of phenolic hydrogens in silylation would leave an appreciable amount of basic ring nitrogens available for association with OPP–OH protons.

The above qualitative correlation of viscosity, calorimetric, and NMR results suggests that, in coal-liquids, asphaltene and pentane-soluble HO fractions are associated intermolecularly through hydrogen bonding involving largely phenolic hydrogens as proton donors. The hydrogen bonding is, in part, responsible for the increase of viscosity of the product oil.

Influence of Hydrogen Bonding on the Viscosity of Coal Liquids

To further investigate the effect of hydrogen bonding involving phenolic OH on the viscosity of coal-liquids, we have compared viscosity data for coal-liquid–amine systems with those for a model system, *o*-cresol–amine mixtures. The results are shown in Figure 4. In systems involving acid-base interaction (—OH---N), the viscosity of the mixtures as a function of *o*-cresol or SRC-II–Acid mole fraction shows a single maximum. Our results are similar to those reported by Felix and Huyskens (*16*) for mixtures of phenol and aliphatic amines, where viscosity, electric conductivity, and volume contraction have been correlated with the size and concentration of ionic proton-transfer complexes ($—O^{-}---H\overset{+}{N}R_3$). No maximum in the viscosity vs. mole fraction curve was observed for the SRC II + Et_3N system because bonded

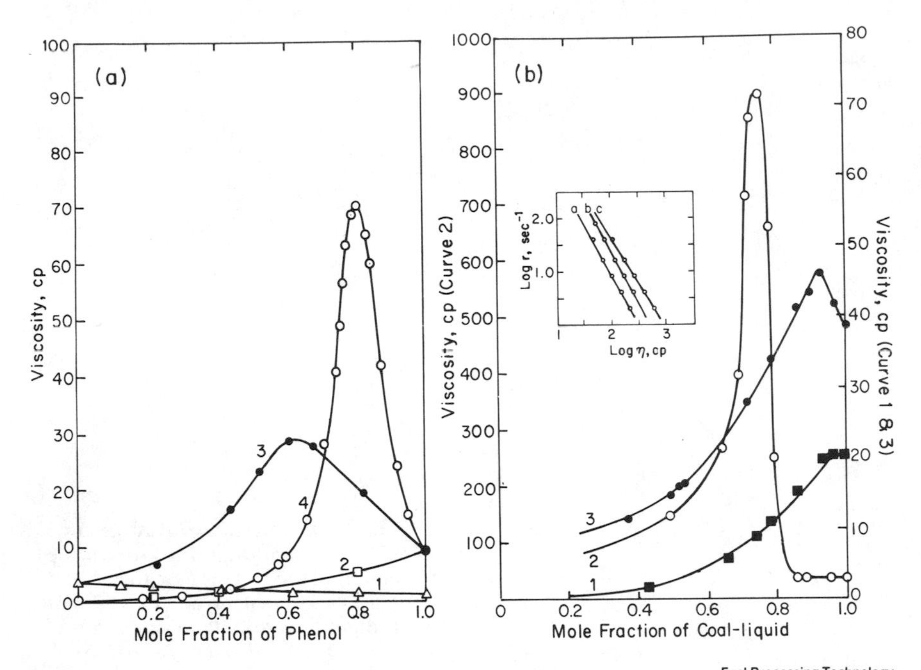

Fuel Processing Technology

Figure 4. (a) Viscosity, in cp, as a function of mole fraction of the phenol at 293 K: (△) anisole and Qu, (□) o-cresol and CCl_4, (●) o-cresol and Qu, (○) o-cresol and Et_3N. (b) Viscosity, in cp, as a function of the mole fraction of the coal liquid at 298 K: (■) SRC II and Et_3N, (○) SRC-II–Acid and Et_3N; (●) SRC-II–Acid and Qu. (Insert) Effect of rate of shear (r) on the viscosity (η) of SRC-II–Acid and Et_3N mixtures. Mole fraction of SRC-II–Acid: (a) 0.491, (b) 0.795, (c) 0.689 (17).

phenolic OH groups already exist in the unfractionated liquid (SRC II). The liquid mixtures show Newtonian behavior over the entire mole fraction region. However, in SRC-II–Acid + Et_3N and SRC-II–Acid + Qu systems, hydrogen bonding involving phenolic OH as proton donor not only defines the viscosity but also the plastic nature of these systems as shown in Figure 4b (insert). Felix and Huyskens (*16*) have shown that the size of the ionic complexes in the mole fraction region below the maximum is larger than that

Table V. Number-Average Molecular Weight of SRC-II–Acid + Et_3N Mixtures (*17*)

Mole Fraction of SRC II-Acid	M_{obs}	M_{calc}[a]
0.677	250	175
0.763	230	184
0.805	220	189
1.000	210	210

[a]Calculated from mole fractions and molecular weights of SRC-II–Acid (M = 210) and Et_3N (M = 101) assuming no complex formation.

Fuel Processing Technology

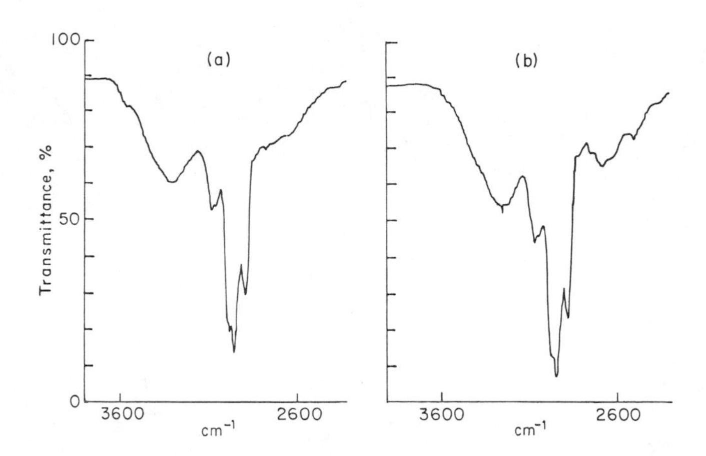

Fuel Processing Technology

Figure 5. Infrared spectra, neat: (a) SRC-II–Acid; (b) SRC-II–Acid and Et_3N (mole fraction of SRC-II–Acid, 0.69) (17).

of the ionic complexes in the mole fraction region above the maximum. The results of our molecular weight determination of SRC-II–Acid + Et_3N mixtures at various mole fractions [Table V (*17*)] qualitatively indicate that the relative size of the molecules is larger at mole fractions below the observed maximum (at mole fraction 0.75) than above the maximum.

The transition from a hydrogen-bonding complex (O—H · · · N) to a proton-transfer complex (O^- · · · $^+$H—N) can be followed by IR spectroscopy. The IR spectra of SRC-II–Acid and SRC-II–Acid + Et_3N mixture at 0.69 mole fraction of SRC-II–Acid are shown in Figure 5. Absorption bands at 2680 cm^{-1} and 2500 cm^{-1}, ascribed to N^+—H · · · O (*18, 19*) and a broad absorption band at 3200–3600 cm^{-1} clearly indicates the presence of both OH · · · N and O^- · · · $^+$H—N species in the SRC-II–Acid + Et_3N system. The self-association of SRC-II–Acid is obvious from the IR spectrum of neat SRC-II–Acid.

Figure 6 shows the IR spectra of CCl_4 solutions of the SRC II + Et_3N and SRC-II–Acid + Et_3N mixtures. Addition of Et_3N to SRC-II–Acid in CCl_4 results in the appearance of strong N^+—H · · · O bands (*18, 19*) at 2630, 2610, and 2500 cm^{-1}. On the other hand, the addition of Et_3N to SRC II shows no significant absorption bands in the region 2500–2800 cm^{-1}. These results indicate that hydrogen bonding involving largely phenolic OH and nitrogen-containing bases yields proton-transfer complexes that are, in part, responsible for the viscosity and non-Newtonian flow of the coal liquids.

Acknowledgments

The authors thank the Department of Energy for support of this work under Contract No. EY-76-S-02-0063.A003 and DE-AC22-80PC 30252. We thank Bradley C. Bockrath for helpful discussions.

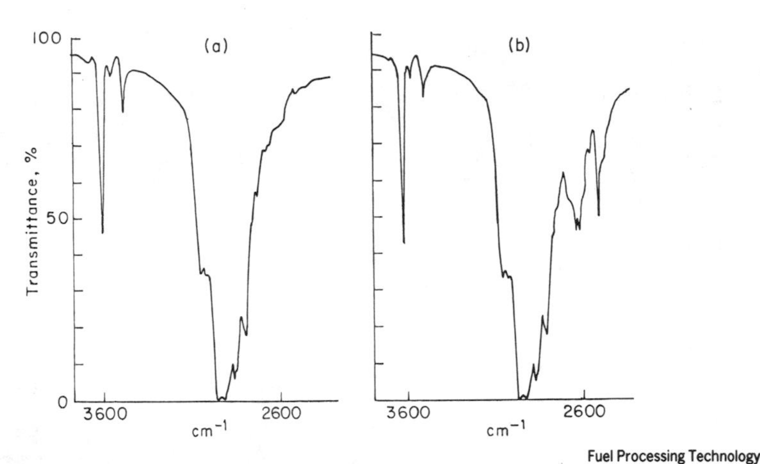

Figure 6. Infrared spectra in CCl_4 solutions, 5-mm cell (KBr): (a) SRC II (0.02M), Et_3N (0.01M); (b) SRC-II–Acid (0.02M), Et_3N (0.01M) (17).

Literature Cited

1. Sternberg, H. W.; Raymond, R.; Akhtar, S. In "Hydrocracking and Hydrotreating," *ACS Symp. Ser.* **1975,** *20,* 111.
2. Sternberg, H. W.; Raymond, R.; Schweighardt, F. K. *Science* **1975** *188,* 49.
3. Burk, E. H.; Kutta, H. W. *Stanford Res. Inst. Coal Chemistry Workshop, Prepr.* (1976) p. 86.
4. Bockrath, B. C.; Lacount, R. B.; Noceti, R. P. *Fuel Process. Technol.* **1978,** *1,* 217.
5. Thomas, M. G.; Granoff, B. *Fuel* **1978,** *57,* 122.
6. Schiller, J. E.; Farnum, B. W.; Sondreal, E. A. *Am. Chem. Soc. Div. Fuel Chem., Prepr.* (Chicago, Aug.–Sept., 1977) *22*(6), 33.
7. Tewari, K. C.; Egan, K. M.; Li, N. C. *Fuel* **1978,** *57,* 712.
8. Tewari, K. C.; Galya, L. G.; Egan, K. M.; Li, N. C. *Fuel* **1978,** *57,* 245.
9. Tewari, K. C.; Kan, N. S.; Susco, D. M.; Li, N. C. *Anal. Chem.* **1979,** *51,* 182.
10. Bolles, T. F.; Drago, R. S. *J. Am. Chem. Soc.* **1965,** *87,* 5015.
11. Yavorsky, P. M. "*PERC Int. Q. Progr. Rep.,*" Apr.–June, 1976; Oct.–Dec., 1976.
12. Brown, J. K.; Ladner, W. R. *Fuel* **1960,** *39,* 87.
13. Tewari, K. C.; Wang, J. T.; Li, N. C.; Yeh, H. J. C. *Fuel* **1979,** *58,* 371.
14. Schwager, I.; Lee, W. C.; Yen, T. F. *Anal. Chem.* **1977,** *49,* 2363.
15. Lee, W. C.; Schwager, I.; Yen, T. F. *Am. Chem. Soc., Div. Fuel Chem., Prepr.* (Anaheim, Mar., 1978) *23*(2), 37.
16. Felix, N. G.; Huyskens, P. L. *J. Phys. Chem.* **1975,** *79,* 2316.
17. Tewari, K. C.; Hara, T.; Young, L-J. S.; Li, N. C. *Fuel Process. Technol.* **1979,** *2303.*
18. Zeegers-Huyskens, Th. *Spectrochim. Acta* **1965,** *21,* 221.
19. Ibid., **1967,** *23A,* 855.

RECEIVED June 23, 1980.

11

Lewis Acids Assisted Degradation of Athabasca Asphaltene

T. IGNASIAK, J. BIMER,[1] N. SAMMAN, D.S. MONTGOMERY, and O.P. STRAUSZ

Hydrocarbon Research Centre, Department of Chemistry, University of Alberta, Edmonton, Alberta, Canada T6G 2G2

Our objective was to determine the extent to which the Lewis acids $AlCl_3$ and $ZnCl_2$ promote degradation of Athabasca asphaltene into lower molecular weight species, and also to identify the chemical changes taking place. In contrast with $ZnCl_2$, $AlCl_3$ degrades asphaltene to some extent at temperatures below 100° C. The solubility of the $AlCl_3$-treated asphaltene depends on the reaction medium, and this is interpreted in terms of the mechanism of typical Friedel–Crafts reactions. At higher temperatures (150°–400° C), $ZnCl_2$ is a more suitable catalyst than $AlCl_3$ for degradation. Below 250° C, $ZnCl_2$ promotes insolubilization of asphaltene in benzene, and this is attributed to the formation of polar groups as a result of cleavage reactions. Above 250° C, conversion into low molecular weight pentane solubles improves. $ZnCl_2$ promotes heteroatom removal, as shown by the decrease in heteroatom content of nonvolatile products with increasing reaction temperature. Hydrogen is essential to compensate for the dehydrogenating effect of $ZnCl_2$ and to obtain optimal pentane solubility.

Asphaltene is a complex, high molecular weight fraction of petroleum. Being a solubility class, it is an extremely heterogeneous mixture, which makes structural investigations on this material very difficult. For this reason, although a great deal of valuable information has been reported, many structural details remain unsettled. The most significant aspects of petroleum asphaltenes are the nature of the carbon skeletal framework, the types of chemical functional groups in which the heteroatoms appear, and the chemical nature of the bridges connecting the subunits of the molecule. These

[1]Current address: Institute of Organic Chemistry of the Polish Academy of Sciences, Warsaw, Poland.

0065-2393/81/0195-0183$05.00/0

structural parameters together will determine the behavior of asphaltene during a given process.

To obtain a more clearly defined picture of these structural features and to establish the relationship between the chemical structure of asphaltene and its reactivity under a variety of conditions, the potential of chemical and thermal degradation reactions as diagnostic tools has been studied. The specific subject of this investigation was the high molecular weight, sulfur rich asphaltene from the Athabasca bitumen.

Degradation, as the term implies, converts a high molecular weight molecule into lower molecular weight species. It is desirable to achieve fragmentation by the selective cleavage of bridging bonds, leaving the backbone intact.

In general, the selectivity of any degradation process depends largely on the specific structure and reactivity of a given compound; however, in the case of asphaltene the multiple structural arrangements present in the polymer drastically reduce the effectiveness of a single reagent in promoting selective degradation. Therefore, the best approach is to critically evaluate data that have been obtained by methods as diverse as possible. We have shown in previous work (*1*) that the molecular weight of Athabasca asphaltene can be significantly decreased during reduction by electron transfer from naphthalene radical anions produced in situ by treatment of naphthalene with potassium in tetrahydrofuran solution and subsequent octylation. The chemical changes effected during this type of reduction are generally considered to be cleavage of ether and thioether bonds, although the cleavage of certain methylene bridges also has been reported (*2*). Since the oxygen present in this asphaltene has been shown to be mostly in the form of hydroxyl groups (*1*, *3*), it was concluded that the reaction taking place in Athabasca asphaltene is largely cleavage of carbon–sulfur bonds. The presence of sulfide linkages has been further documented by studies on the thermal degradation of the Athabasca asphaltene in the presence of tetralin (*4*). The decrease in molecular weight of the acetylated, methylated, and silylated asphaltene, along with auxiliary IR evidence, suggested that the hydroxyl groups are strongly involved in intermolecular hydrogen bonding.

From these results it was concluded that sulfide linkages and oxygen-based hydrogen bonding play important roles in the molecular size of asphaltene. The presence of the relatively weak sulfide bonds may explain the relative ease with which asphaltene can be partly depolymerized under moderate thermal conditions where extensive cracking and coke formation can be excluded (*5*). The reductive systems involving sodium or lithium in liquid ammonia appeared to be less effective in degrading asphaltene, as reflected by only a moderate decrease in the molecular weight (*6*, *7*). Model reactions showed that in these cases certain sulfides, for example, dialiphatics, do not react.

However, the lithium-reduced asphaltene exhibits an unusual thermal reactivity resulting in a higher yield of lower molecular weight pentane-soluble fraction and higher degree of desulfurization, as compared with the original asphaltene (*5*). Apparently this is attributable to the cleavage of the carbon–sulfur bonds. Moreover, the introduction of hydrogen into the asphaltene molecule during reduction leads to saturation of the free radicals produced during thermolysis, thereby preventing repolymerization. The enhanced desulfurization occurring on thermal treatment of the reduced asphaltene may be related to the formation of thermally unstable thiols during reduction.

The pentane-soluble portion of asphaltene thermolyzed at 300° C has been found to contain a full complement of *n*-alkanes, acyclic isoprenoids, small ring saturated and small ring aromatic compounds, typical of unbiodegraded conventional crude oil (*8*). However, in all cases only a small fraction of the degraded asphaltene can be analyzed in this manner, and therefore the question of the size and homogeneity of the remaining hydrocarbon units remains unresolved.

The results of the depolymerization studies such as metal reductions and low temperature solvolysis have shown that the mechanism and the rate of degradation of asphaltene depend to a large extent on the chemical environment.

It is reasonable to assume that the presence of a catalyst may accelerate the scission of given bonds, and thereby reduce the temperature requirement for degradation to the point at which secondary reactions leading to extensive cracking and polymerization are minimized.

Typical Lewis acids such as $AlCl_3$ and $ZnCl_2$ display excellent catalytic activity in most Friedel–Crafts type reactions (*9, 10*) and are also known to catalyze coal liquefaction and solubilization (*11–26*). Because of this latter property, acid-catalyzed depolymerization has become a useful method for structural investigation of coal (*27*). The rationale for solubilization is interpreted either in terms of depolymerization via rupture of the bridges or, since the overall reaction in general is that of Friedel–Crafts alkylation or acetylation, in terms of the effect of the sidechains introduced on the aromatic ring system.

In this study the effects of the Lewis acids $AlCl_3$ and $ZnCl_2$ on the degradation of Athabasca asphaltene have been investigated under various conditions including temperature, ranging from 40° to 400° C, and pressures up to 1800 psi of hydrogen or nitrogen. Our objective was to determine the extent to which these catalysts promote degradation into lower molecular weight species and also to identify the chemical changes taking place, to complement our previous structural investigations. In particular it was desirable to establish those conditions that improve the selectivity with which various bonds are cleaved.

Experimental

Materials. Athabasca asphaltene was prepared from an oil sand sample from the GCOS (now Suncor Inc.) quarry according to a standard procedure used in this laboratory (*9*). Solvents were refluxed over CaH_2 and redistilled. Nitromethane was distilled and stored over molecular sieves. $AlCl_3$ was sublimed and ground under nitrogen; $ZnCl_2$ was dried at 110° C in a vacuum oven.

Experiments Under Reflux. To a solution of asphaltene (2 g) dissolved in 100 mL of the required solvent, $AlCl_3$ (2 g) was added and the reaction mixture was magnetically stirred for 17 h under nitrogen at reflux temperature. When nitromethane was used as the solvent, $AlCl_3$ was dissolved in the solvent and added to a suspension of asphaltene in CH_3NO_2, and the reaction temperature was set to 70° C. After cooling the reaction mixture, 100 mL 1*N* HCl was added and stirring was continued for 1 h. When a volatile solvent such as benzene or CH_2Cl_2 was used, it was removed in a Rotavapor. The resulting precipitate was filtered, washed extensively with water followed by a small volume of acetone, and dried first in a vacuum oven overnight at 60° C and then in a desiccator under high vacuum. When benzene was the solvent, the precipitate was very sticky, which made filtration extremely difficult. It was therefore flushed with acetone and dissolved in CH_2Cl_2. The CH_2Cl_2 solution was dried over anhydrous Na_2SO_4 and centrifuged for 30 min at 2500 rpm, to remove all inorganic matter. The solvent was then removed and the residue was dried overnight in a desiccator under high vacuum.

Experiments in a Batch System. In the experiments with molten catalyst, asphaltene and catalyst were mechanically mixed and placed in a reactor. In the experiments using a solvent, the catalyst was first impregnated on asphaltene and suspended in the solvent.

To impregnate asphaltene, the catalyst was dissolved in diethyl ether and the solution added to a solution of asphaltene in benzene. Solvents from the resulting suspension were removed in a Rotavapor and the residue was dried at 110° C in a vacuum oven.

Reactions were conducted in a 300-cm^3, Magne Dash stirred autoclave equipped with a fitted glass liner. After the reactants were introduced, the system was flushed several times with a given gas, which was then adjusted to the required initial pressure. The valves were then closed, thus sealing the autoclave. The temperature sequence was a 30–45 min heating period to the desired reaction temperature, followed by a cooling period of about 3 h. The asphaltene-to-catalyst ratio was 1:1 by weight, and the actual proportions used were as follows:

1. 1 g each of asphaltene and $AlCl_3$ in 50 mL benzene.
2. 2 g each of asphaltene and $ZnCl_2$ in 70 mL solvent.
3. 3 g each of asphaltene and molten catalyst.

The reaction mixtures were treated with 1*N* HCl and left overnight to allow the catalyst and metal complexes to decompose. The solvents were then distilled off and the precipitate was filtered, washed exhaustively with 1*N* HCl, followed by 10% $NaHCO_3$ and water. The precipitate was then dried at 60° C in vacuo.

The crude products (nonvolatiles) were then Soxhlet extracted to yield the pentane-soluble fraction, the benzene-soluble fraction, and insoluble residue.

The yield of gases was calculated from the weight difference between substrate and product.

Hydrocarbon Class Analysis. The hydrocarbon class or type separation of the pentane-soluble fraction was performed on a silica gel–alumina column according to system II of Sawatzky et al. (28) with the following modifications: the Pyrex glass column, 6 mm i.d. × 50 cm long, fitted with a ground glass joint at the top, was packed in the upper half with a 4 g Merck silica gel 60 (70–230 mesh), activated at 250° C > 12 h, and the lower half packed with 5 g Woelm W200 alumina (activity grade super 1). The column was prewetted with pentane, then a 100-mg sample in 1 mL pentane was admitted onto the column. The eluents were forced downwards through the column as a result of the pressure exerted by weights attached to the barrel of a 50-mL syringe connected to the column via a Luer Lok needle and embedded in a Teflon plug fitting the column joint. The elution rate was kept constant at about 1 mL/min by varying the mass of the weights as necessary. The following solvents were then used for consecutive elution of the sample charges:

Eluent	*Volume (mL)*
Pentane	25
5% Benzene/pentane	30
15% Benzene/pentane	30
Benzene	15

Elution was then terminated, the column packing removed and Soxhlet extracted with benzene/methanol (60:40) to retrieve the residual portion.

Results and Discussion

Reaction with Aluminum Chloride. Asphaltene reacts readily with $AlCl_3$ even at temperatures well below 100° C. The characteristics of the product, mainly in terms of solubility, are strongly influenced by the reaction medium (Table I). Asphaltene treated with $AlCl_3$ in benzene remains fully soluble, while when the remaining solvents are used as a reaction medium the solubility drops to a mere 10%–20%.

The effect of solvent on the solubility of the product can be explained on the basis of the general mechanism of a typical Friedel–Crafts reaction. Thus, asphaltene reacts with $AlCl_3$ to form intermediate carbonium ions, which then undergo electrophilic substitution. If substitution occurs within the asphaltene molecule new bonds are formed and, depending on the size of initial fragments, the molecule may grow bigger and, thus, less soluble.

Table I. Characteristics of the Reaction Product of Asphaltene with $AlCl_3$

Reaction Medium	*Yield (%)*	*Solubility in CH_2Cl_2 (%)*	*H/C*	*Cl (%)*	*N (%)*
Benzene	106	99	1.14	1.9	
Dichloromethane	102	23	1.09	1.1	
Nitromethane	103	12	1.15	1.6	1.5
Nitromethane/benzene	102	16	1.12	1.1	1.4
Asphaltene		100	1.22		1.1

However, benzene is a strong nucleophile and is very likely to react with the intermediate carbocations, thus preventing them from undergoing intramolecular interactions. Nitromethane, on the other hand, is a relatively inert solvent and the decrease in solubility indicates that internal polymerization prevails. Dichloromethane itself reacts in the presence of $AlCl_3$, and may take part in bifunctional alkylation of asphaltene fragments with the formation of new methylene bridges, thus increasing the molecular size and lowering the solubility. The use of a nitromethane–benzene mixture was intended to approximate the fully homogeneous reaction and, therefore, the poor solubility of the product in this case is difficult to explain. In each case, the H/C ratio is always lower than that of the substrate. The chlorine content does not exceed 2%.

In benzene, small amounts of $AlCl_3$ (about 10% by weight of asphaltene) do not effect degradation and for reaction to occur, high concentrations of $AlCl_3$ (1:1 by weight) are required. The requirement for excessive amounts of catalyst seems to be a result of poor contact between asphaltene and $AlCl_3$, attributable to unusual properties of the asphaltene–$AlCl_3$ complex that make it precipitate out of the solution, thus trapping most of the unreacted catalyst. The characteristics of the product are given in Table II.

The product from the reaction carried out in benzene shows a decrease in the average molecular weight, indicating that degradation has taken place. This was further confirmed by preparative GPC separation on Bio-Beads SX-1. The results, shown in Table III, show that the concentrations of the two highest molecular weight fractions of asphaltene, their sum comprising approximately 69% by weight vs. 45% in $AlCl_3$-treated asphaltene, have been reduced by one third. The lower H/C ratio for the treated asphaltene is accompanied by an increase in the amount of aromatic hydrogen, as calculated from the integration curve of the ^{1}H NMR spectrum. The same reaction carried out in deuterated benzene showed that this increase comes from incorporation of phenyl groups from the solvent. The number of phenyl groups introduced, calculated from the carbon content before and after

Table II. Reaction of Asphaltene with $AlCl_3$ in Benzene

Characteristics	*Treated Asphaltene*
Molecular weight (in benzene)	1200
H/C	1.14
Sulfur (%)	6.93
Extractability (wt %)	
pentane	22
benzene	78
H_{ar} (%)	15
Per 100 C atoms	
phenyl groups added	1.2
hydrogen atoms (loss)	~4

Table III. GPC Separation of the $AlCl_3$-Treated Asphaltene

	Percent Distribution	
Fraction Molecular Weight	*Asphaltene*	*Product*
>6000	43	26
4000–6000	26	19
	69	45

reaction and also from the aromatic hydrogen increase, amounts to 1.2 groups per 100 carbon atoms or 1 group per molecule of reaction product. The hydrogen balance showed a deficit of 4 hydrogen atoms per 100 carbons. In other words, phenylation is accompanied by dehydrogenation.

As a result of treatment with $AlCl_3$ in benzene, more than 20% of asphaltene became soluble in pentane. The differences in the chemical composition of the pentane and benzene fractions are shown in Table IV.

The pentane-soluble fraction has a low molecular weight, a higher H/C ratio compared with the original asphaltene, lower sulfur content, and very low nitrogen content. Integration of the ^1H NMR spectrum suggests that about 5 phenyl groups are attached per 100 carbon atoms.

Although polar materials constitute the major portion of the pentane solubles, saturates and aromatics are also found in appreciable concentrations (Table V). The relatively low H/C ratio of the saturate fraction suggests a high concentration of saturated cyclics or olefins. Thermogravimetric analysis indicated that only about 40% of the pentane solubles is volatile under typical gas chromatographic conditions.

Under the same conditions, experiments carried out with model compounds showed that benzyl sulfide underwent 100% conversion, the major product being diphenylmethane (from benzyl carbonium ion attack on benzene), whereas only 23% and 26% conversion to mixtures of unidentified compounds took place for *n*-heptyl sulfide and bibenzyl, respectively.

The effects of temperature, solvent, and pressure of added N_2 or H_2 are listed in Table VI.

Table IV. Chemical Composition of the Pentane and Benzene Fractions of $AlCl_3$-Treated Asphaltene

	Fraction	
Characteristics	*Pentane*	*Benzene*
Molecular weight (in benzene)	380	2400
H/C	1.33	1.10
Sulfur (%)	4.58	7.44
Nitrogen (%)	0.10	1.03
H_{ar} (%)	21	n.d.[a]
Phenyl groups per 100 C atoms	5.1	n.d.[a]

[a] Not determined.

Table V. Hydrocarbon Type Class Separation of the Pentane-Soluble Fraction

Fraction	*Yield*[a]	*Molecular Weight*[b]	*H/C*	*Percent*	
				N	*S*
Saturates	15	340	1.66	0.0	0.1
Aromatics					
mono-	15	260[c]	1.17	0.0	2.6
di-	8				
poly-	11	540	1.20	0.2	6.3
Polars	51	630	1.34	0.3	6.0

[a]% of the pentane solubles.
[b]In benzene.
[c]For the combined mono- and diaromatic fractions.

Increasing the temperature appeared to be of no advantage. The surprisingly low H/C ratio of the product from the reaction carried out in cyclohexane points to advanced dehydrogenation. Moreover, blank experiments indicated that $AlCl_3$ promotes the decomposition of these solvents with the formation of a variety of products ranging from di- and triaromatics to insoluble polymers. This results in an unreasonably high weight increase of the treated asphaltene.

To avoid undesirable side effects resulting from the solvent reactivity, dry asphaltene was treated with $AlCl_3$ melt, that is, at a temperature slightly above the melting point of $AlCl_3$ (220° C) under hydrogen pressure. The results, shown in Figure 1, illustrate the general trend towards formation of gases and insoluble products; the yield of the former increases rapidly with the time of reaction. The reaction, in terms of product yields and heteroatoms removed, is

Table VI. Reaction with $AlCl_3$ in a Pressurized System

	Experiment Number	
	1	*2*
Conditions		
temperature (°C)	160	300
solvent	benzene	cyclohexane
gas pressure (psig)	200 (N_2)	850 (H_2)
Yield (% of substrate)		
solid product	127	158
Extractability (% of product)		
pentane fraction	28	13
benzene fraction	44	32
residue	28	55
H/C of solid product	0.96	0.83

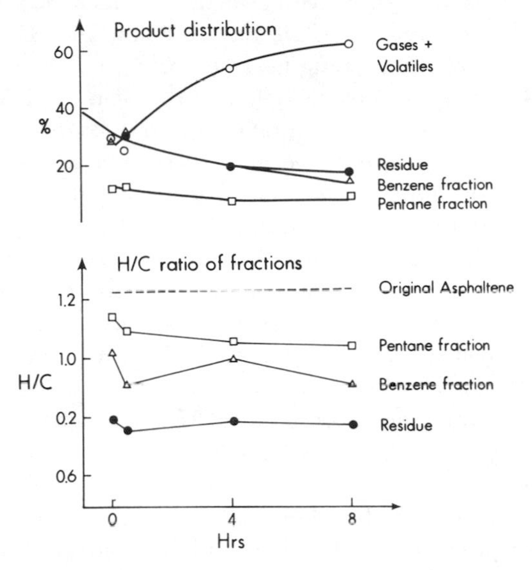

Figure 1. Degradation of asphaltene with molten $AlCl_3$

more or less complete in four hours. The H/C ratios of the fractions do not change with reaction time, but they are distinctly lower than that of the original asphaltene.

The destructive decomposition of asphaltene that eventually leads to high conversion into gases may arise from improper contact between the solid–liquid–gas phases. The role of hydrogen is to hydrogenate and stabilize the fragments produced upon catalyzed bond scission. However, since the system discussed above was not stirred, access of hydrogen into the asphaltene molecule might have been limited by the layer of melted catalyst surrounding the substrate, which would then be completely vulnerable to the high cracking activity of $AlCl_3$.

It is evident that, depending on the conditions employed, $AlCl_3$—one of the strongest Lewis acids—can promote many secondary reactions involving the primary products formed during asphaltene degradation. Thus, in addition to initial depolymerization as a result of sulfide and methylene bond cleavages, inter- and intramolecular alkylation, dehydrogenation of hydroaromatic structures, aromatic condensation, and indiscriminate cracking all may occur. The overall result is that the actual extent of degradation into condensible low molecular weight species is insignificant.

Reaction with Zinc Chloride. In contrast to other Lewis acids, $ZnCl_2$ is a relatively weak catalyst for hydrogenation and hydrocracking of single ring aromatics (*17*, *29*), and since it has been reported (*30*) to be far more selective in the cleavage of certain bonds than is $AlCl_3$, it would appear to be a very attractive reagent from the structural point of view.

Since $ZnCl_2$ does not react with asphaltene under the same conditions as those employed in the case of $AlCl_3$, that is, below 100° C, the reaction was investigated at temperatures ranging from 150° to 400° C in an autoclave. The merit of the batch autoclave system is that it is possible to conduct the reaction under various gaseous atmospheres and pressures and, in addition, it permits a variety of solvents to be employed to control chain propagation reactions. To obtain optimal conversion of asphaltene into pentane-soluble materials, the presence of a solvent appeared to be essential to improve the dispersion of $ZnCl_2$ and facilitate agitation; otherwise, in the absence of solvent, the interaction of asphaltene with $ZnCl_2$ melt at 300° C led to advanced decomposition to gaseous products similar to the situation observed when the $AlCl_3$ melt was used. However, $ZnCl_2$ appeared to be significantly more efficient than $AlCl_3$ in converting heteroatoms into gaseous products, the best examples of which are illustrated in Figure 2, where it is seen that an appreciable though slow denitrogenation and rapid and substantial desulfurization take place with time.

The disadvantage of the closed system is that low molecular weight primary scission products, attributable to a long residence time, may undergo secondary cracking and polymerization reactions. In addition, some of the low molecular weight oils may be lost upon evaporation of the solvent. Therefore, in this type of experiment, emphasis is placed on the overall product

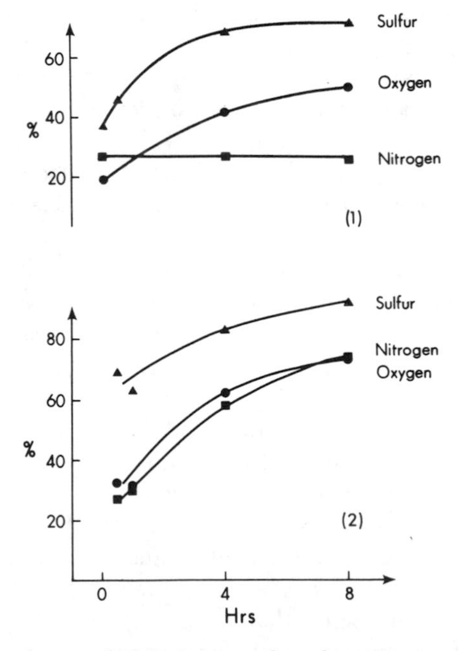

Figure 2. Degradation of asphaltene with molten Lewis acids (H_2 850 psig); heteroatom removal: (1) $AlCl_3$ (220° C); (2) $ZnCl_2$ (300° C)

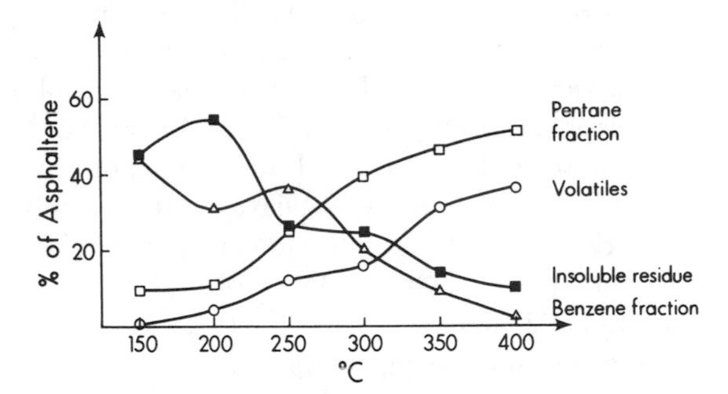

Figure 3. Effect of temperature on the product distribution with $ZnCl_2$ catalyst

distribution and the behavior of high molecular weight components, while the chemical nature of the low molecular weight materials is more closely evaluated in a parallel series of experiments carried out in a flow system (*31*). The latter appears to be more suitable for mechanistic and structural studies, since most of the volatile products can be carried away from the reaction zone in the carrier gas.

Although benzene is a better solvent for asphaltene than cyclohexane, and the product distribution obtained from the hydrogenation of Athabasca asphaltene in benzene and cyclohexane is almost the same, experiments with model compounds showed that carbonium ions formed by ether and sulfide bond cleavages react with benzene. Thus, cyclohexane, being less reactive towards $ZnCl_2$, is the solvent of choice. It was also noted that the effectiveness of the catalyst in promoting degradation of asphaltene depends upon the manner in which the catalyst is introduced to the solution. The most satisfactory method, therefore applied in this study, appeared to be that of impregnating asphaltene with a $ZnCl_2$ solution. The effect of temperature on the product distribution obtained from the reaction of $ZnCl_2$ with asphaltene in cyclohexane under a hydrogen pressure of 850 psig is illustrated in Figure 3. The yields of volatile products and of the pentane-soluble fraction increase with temperature, up to 37% and 51%, respectively. The yield of the benzene-insoluble fraction is substantial at lower temperatures, reaches a maximum at 200° C, and then gradually decreases to about 10% at 400° C.

The product distribution indicates that a certain specific reaction between asphaltene and $ZnCl_2$ takes place at quite low temperatures (150°–200° C) and results in substantial insolubilization of treated asphaltene; for example, at 200° C about 55% of asphaltene becomes insoluble in benzene. The following possibilities can account for this insolubilization:

1. Formation of salts or complexes with $ZnCl_2$.
2. Increase in polarity attributable to the formation of new functional groups as a result of cleavage of sulfides or ethers.
3. Formation of new covalent bonds by alkylation or homolytic recombination.

Although the benzene-insoluble residue at 200° C contained 1.41% zinc (determined by atomic absorption), no correlation between the zinc content in various reaction products and the solubility in benzene was observed. Moreover, severe and prolonged treatment with 1*N* HCl would certainly decompose the complex and restore the solubility, but this was not observed.

Between 150° and 200° C homolytic cleavage and recombination of the resulting fragments are less likely to occur. However, in the presence of Lewis acids, some sulfide and ether bridges may be involved in ionic reactions. It has been recently reported (32) that a variety of ethers containing at least one methylene group adjacent to the oxygen atom can be cleaved easily by $ZnCl_2$ at 300° C. Reactions on model compounds at 200° C gave similar results for ethers and sulfides and indicated that the carbonium ions formed by ether or sulfide bond cleavage may react with any aromatic system via inter- or intramolecular alkylation (e.g., the major products from the reaction of benzyl ether with $ZnCl_2$ were diphenylmethane when benzene was used as solvent, and insoluble polymer when cyclohexane was the solvent).

Similar mechanisms resulting in the formation of new functional groups and simultaneous intramolecular alkylation with or without changing the molecular weight also can be readily envisaged as occurring in a complex asphaltene molecule:

$$\text{Ar–S–R} \xrightarrow{ZnCl_2} \begin{cases} \text{Ar(SH)(R)} \\ \text{Ar–SH} \quad (-\,R^-) \end{cases}$$

When the benzene-insoluble residue obtained at 200° C was subjected to nonreductive alkylation (33) the solubility was restored to 80%, and this strongly implies that the generation of new functional groups is responsible for the lower solubility at this temperature. However, in view of inter- or intramolecular alkylations that may accompany the cleavage reactions, it is not possible to correlate the extent of degradation with changes in the molecular weights.

With increasing temperature the amount of insolubles decreases, and this corresponds to a depletion of the functional groups, as reflected by increasing losses in sulfur and oxygen contents. Indeed, a significant feature of the $ZnCl_2$-catalyzed degradation of asphaltene is the amount of heteroatoms removed from the recovered product as a function of temperature (Figure 4).

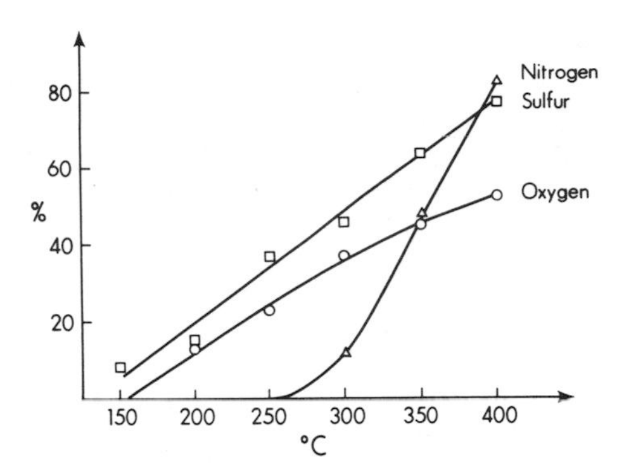

Figure 4. Effect of temperature on the removal of heteroatoms from nonvolatile products

Oxygen and sulfur removal already takes place below 200° C and their rates are almost linear with increasing temperature, reaching about 53% and 78%, respectively, at 400° C. Most of the nitrogen is removed between 350°–400° C. Above 300° C there is a slight decrease in the H/C ratio, a substantial decrease in the proportions of aliphatic to aromatic hydrogens (Table VII) and an almost constant concentration of polar compounds in the pentane-soluble fraction.

The effect of temperature on the formation of hydrocarbons, as revealed by class separation of the pentane-soluble fraction, is shown in Figure 5.

To assess the catalytic action of $ZnCl_2$ on the degradation of asphaltene, the results are compared with those obtained in the absence of $ZnCl_2$. All the reactions were carried out in cyclohexane and in the presence of either nitrogen or hydrogen at 300° and 400° C. The results are summarized in Tables VIII and IX. The reaction temperature of 300° C was chosen to minimize carbon–carbon bond scission and yet bring about the cleavage of bonds involving sulfur and oxygen.

Table VII. Effect of Temperature on the Characteristics of the Pentane-Soluble Fraction[a]

Temperature (°C)	*Yield (%)*	*Molecular Weight*	*H/C*	H_{aliph}/H_{ar}
150	9.6	840	1.36	n.d.[b]
200	10.8	770	1.37	12.0
250	25.1	490	1.36	9.8
300	39.3	430	1.37	8.8
350	46.1	390	1.31	8.4
400	51.2	250	1.30	5.2

[a]Conditions: catalyst, $ZnCl_2$; catalyst asphaltene weight ratio, 1:1; solvent, cyclohexane; H_2 pressure, 850 psig; duration, 1 h.
[b]Not determined.

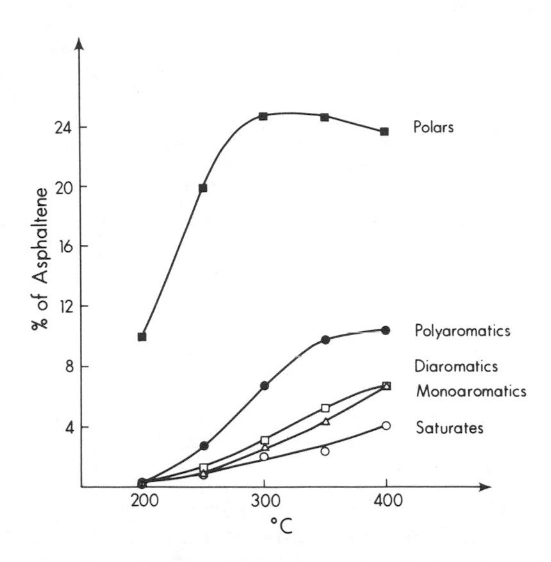

Figure 5. Effect of temperature on the yields of hydrocarbons

At 300° C, thermolysis of asphaltene dissolved in cyclohexane gives results similar to those previously described (5). Under the same conditions $ZnCl_2$ accelerates the degradation reactions, the main quantitative difference being higher yields of gaseous and benzene insoluble products as well as a marked increase in the concentrations of saturates, mono-, di-, and polyaromatics in the pentane-soluble fraction. The yields of these compound classes were twice as high as those obtained in the absence of $ZnCl_2$. The low H/C ratio of the recovered solid product points to the dehydrogenating properties of $ZnCl_2$. Indeed, bituminous coal has been reported (*34*) to evolve hydrogen in the presence of $ZnCl_2$ under an inert atmosphere even at 200° C; this was attributed to dehydrogenation of hydroaromatic structures.

Replacing nitrogen by hydrogen substantially improves the pentane solubility and simultaneously suppresses the formation of insolubles. Hydrogen also helps to compensate for the dehydrogenating action of $ZnCl_2$. The yields of saturated and aromatic hydrocarbons released during degradation are not affected by the nature of the gas. However, the stabilizing role of hydrogen is reflected by higher yields of lower molecular weight, pentane-soluble, polar materials resulting from the cleavage reactions. Increasing the hydrogen pressure from 200 psi to 1800 psi effects a slight increase in the yields of pentane solubles (from 33% to 41%), volatiles (from 18% to 24%), and hydrocarbons (from 14% to 19%). However, the chemical characteristics of the products (e.g., H/C, H_{aliph}/H_{ar}, mw) remain unchanged. The dominant reaction governing the overall kinetics is the formation of intermediate ions by bond scission (sulfide, ether, dealkylation), and the subsequent reactions with hydrogen are less important.

Table VIII. Product Distribution from the Catalyzed and Noncatalyzed Degradation of Asphaltene

		Distribution (% of asphaltene)								
							Aromatics			
$ZnCl_2$	Gas	Pentane Fraction	Benzene Fraction	Insoluble Residue	Volatiles (by difference)	Saturates	Mono-	Di-	Poly-	Polars
300° C										
no	N_2[a]	20.9	66.7	0.0	12.4	0.8	0.8	0.7	1.2	17.4
yes	N_2[a]	23.5	10.4	42.1	24.0	1.4	2.7	3.4	5.5	10.5
yes	H_2[b]	39.3	20.0	25.1	15.6	2.0	2.6	3.1	6.7	24.9
400° C										
no	N_2[a]	31.4	5.7	34.8	28.1	3.7	3.3	2.9	4.7	16.8
no	H_2[b]	50.6	20.0	9.6	19.8	5.6	3.6	3.2	6.4	31.8
yes	N_2[a]	34.9	0.8	58.5	5.8	2.1	5.4	4.9	7.3	15.2
yes	H_2[b]	51.2	2.2	10.3	36.3	4.0	6.6	6.6	10.3	23.7

[a] 15 psig.
[b] 850 psig.

Table IX. Product Characteristics from the Catalyzed and Noncatalyzed Degradation of Asphaltene

		Pentane Fraction					*Benzene Fraction*					*Insoluble Residue*				*Total Solid Product*		
				Percent					*Percent*				*Percent*					
$ZnCl_2$	*Gas*	*Molecular Weight*	*H/C*	*S*	*N*	*Cl*	*Molecular Weight*	*H/C*	*S*	*N*	*Cl*	*H/C*	*S*	*N*	*Cl*	*H/C*	*S*	*N*
300° C																		
no	N_2[a]	650	1.38	7.4	0.5	0.3	5300	1.14	7.2	1.3	1.1	—	—	—	—	1.20	7.3	1.1
yes	N_2[a]	390	1.33	4.5	0.2	0.1	1300	1.05	5.8	0.9	0.8	0.77	4.3	1.5	4.7	0.98	4.6	1.0
yes	H_2[b]	430	1.37	4.8	0.2	1.0	1700	1.10	5.3	1.3	1.9	0.88	4.5	1.7	5.0	1.16	4.8	0.9
400° C																		
no	N_2[a]	410	1.39	6.4	0.5	0.8	1100	0.98	7.3	1.5	1.5	0.66	7.2	1.8	1.1	1.01	6.9	1.2
no	H_2[b]	410	1.36	6.1	0.8	—	2200	0.89	5.9	1.9	1.3	0.80	7.0	1.5	1.8	1.18	6.2	1.2
yes	N_2[a]	300	1.24	3.2	0.2	—	—	—	—	—	—	0.52	1.7	1.0	0.7	0.78	2.3	0.7
yes	H_2[a]	250	1.30	2.0	0.1	0.2	970	0.83	2.3	1.3	0.6	0.55	4.3	0.6	1.0	1.17	2.4	0.2

[a] 15 psig.
[b] 850 psig.

The negligible effect of hydrogen pressure on the degradation of asphaltene is not really too surprising. It has been reported (*18*, *21*, *23*) that hydrogen pressure has a strong influence on the $ZnCl_2$-catalyzed hydrocracking of coal and SRC, but the reactions were carried out at higher temperatures, usually above 400° C. Zielke et al. (*18*) noted that the role of hydrogen pressure is substantially diminished at lower temperatures.

At 400° C, with a faster rate of thermal decomposition, the role of hydrogen becomes more important in stabilizing the intermediate asphaltene fragments and, thus, preventing polymerization and dehydrogenation in either the catalytic or the noncatalytic degradation mode of treatment. Hydrogen also improves the yields of hydrocarbons. Although the proportions of pentane solubles do not seem to be affected by the catalyst, in the presence of $ZnCl_2$ the yields of mono- and diaromatics are about 70% higher for the catalytic as compared with the noncatalytic degradation.

However, at both temperatures the presence of $ZnCl_2$ significantly improves the quality of the pentane-soluble products in terms of low molecular weight and low nitrogen and sulfur contents.

The gas chromatograms of the hydrocarbon fractions indicate that asphaltene consists of complex macromolecules that decompose to yield a wide distribution (from $\simeq C_{10}$ to $\simeq C_{35}$) of molecules within each of the saturate, mono-, di-, and polyaromatic and polar classes.

Conclusions

Asphaltene reacts with $AlCl_3$ even at temperatures below 100° C. The intermediate carbonium ions formed by the catalyst action on asphaltene tend to undergo intraelectrophilic substitution, leading to polymerization and loss of solubility. A nucleophilic solvent such as benzene suppresses these intramolecular interactions, and the mw of the phenylated product is lower than that of original asphaltene, indicating that degradation has taken place. Reactions with model compounds showed that, below 100° C, the extent of cleavage of carbon–sulfur bonds or methylene bridges is limited. On the other hand, the use of $AlCl_3$ at higher temperatures (150°–300° C) appeared to be of no advantage as far as the degradation into condensable low molecular weight species was concerned; on the contrary, polymerization, dehydrogenation, and advanced cracking to gases prevailed.

The reaction of $AlCl_3$ with benzene and cyclohexane at 160° and 300° C, respectively, gave a wide variety of products that greatly complicated the interpretation of the results when asphaltene dissolved in these solvents was heated with $AlCl_3$. The complexity of the reactions catalyzed by $AlCl_3$ suggested that a milder acid catalyst such as $ZnCl_2$ be employed.

As expected, $ZnCl_2$ was a more suitable catalyst than $AlCl_3$ in the degradation studies carried out at temperatures above 100° C. At lower temperatures (150°–200° C) certain cleavage reactions occur, leading to the formation of polar functional groups that substantially decrease the product solubility in benzene. With increasing temperature, as a result of more extensive interaction of $ZnCl_2$ with heteroatoms, the sulfur, oxygen, and nitrogen contents of the products decrease and the rate of conversion into lower molecular weight pentane-soluble and volatile products increases. This also results in an increase in the saturate, mono-, di-, and polyaromatic classes, the yields of which at given temperature are always higher than those obtained from purely thermal degradation.

As far as structural investigations are concerned, the choice of a 300° C reaction temperature appears to be correct, since at this temperature thermal carbon–carbon bond scission has been shown previously to be minimal and the yield of the low molecular weight soluble fraction is relatively high. At 400° C, thermal degradation of the asphaltene is extensive. However, the catalytically degraded, pentane-soluble portion of asphaltene, as compared with that from the noncatalytic degradation (both at 400° C) is characterized by its superior quality in terms of low sulfur and nitrogen contents and low molecular weight.

The presence of hydrogen is essential to counteract the dehydrogenating effect of $ZnCl_2$ and also to obtain optimal pentane solubility.

The results of the degradative methods so far employed all point to a great degree of chemical complexity of Athabasca asphaltene and clearly show that it is not possible to envision an asphaltene structure based on a single building unit. On the other hand, because of the presence of connecting bridges, of which sulfides are one example, mild degradation can produce appreciable yields of low molecular weight, pentane-soluble polar materials that, because of the relatively mild conditions employed, probably represent the actual structural units of asphaltene.

Acknowledgments

This work was supported by the Alberta Oil Sands Technology and Research Authority. We thank Dr. E.M. Lown for reading the manuscript.

Literature Cited

1. Ignasiak, T.; Kemp–Jones, A. V.; Strausz, O. P. *J. Org. Chem.* **1977,** *42,* 312.
2. Lazarov, L.; Angelov, S. *Fuel* **1980,** *59,* 55.
3. Ignasiak, T.; Strausz, O. P.; Montgomery, D. S. *Fuel* **1977,** *56,* 359.
4. Ignasiak, T.; Strausz, O. P. *Fuel* **1978,** *57,* 617.
5. Ignasiak, T.; Ruo, T. C. S.; Strausz, O. P. *Am. Chem. Soc., Div. Fuel Chem., Prepr.* (Honolulu, Apr., 1979) *24*(2), 178.
6. Ignasiak, T.; Strausz, O. P. Presented at the *Alberta Sulfur Res. Ltd. Sulfur Symp. 27th Canadian Chem. Eng. Conf. Sulfur Week, Calgary, Alberta, Oct. 23–27, 1977.*
7. "The Molecular Structure and Chemistry of Alberta Oil Sand Asphaltene," AOSTRA/University Agreement #30, Final Report, Dec., 1979.
8. Rubinstein, I.; Spyckerelle, C.; Strausz, O. P. *Geochim. Cosmochim. Acta* **1979,** *43,* 1.
9. Galloway, N. O. *Chem. Rev.* **1935,** *17,* 376.
10. Olah, G. A. "Friedel–Crafts Chemistry"; John Wiley & Sons: New York, 1973.

11. Storch, H. H. *Ind. Eng. Chem.* **1937,** *29,* 1367.
12. Weisser, O.; Landa, S. "Sulphide Catalysts, their Properties and Applications"; Friedrich Vieweg & Sohn: Braunschweig, West Germany, 1972.
13. Weller, S.; Pelipetz, M. G. *Ind. Eng. Chem.* **1951,** *43,* 1243.
14. Storch, H. H. In "Chemistry of Coal Utilization"; Lowry, H. H., Ed.; John Wiley & Sons: New York, 1945; Vol. 2, pp. 1750–1796.
15. Weller, S. "Catalysis"; Emmett, P. H., Ed.,; Reinhold: New York, 1956; Vol. 4, Chap 7, pp. 513–527.
16. Faingold, S. I.; Vallas, K. R. *Iz. Akad. Nauk. Est. SSR, Ser. Tekh. Fiz.-Mat. Nauk.* **1957,** *6,* 245.
17. Zielke, C. W.; Struck, R. T.; Evans, N.M.; Costanza, C. P.; Gorin, E. *I & EC Process Des. Dev.*, **1966,** *5,* 151.
18. Ibid., 158.
19. Kawa, W.; Feldmann, H. F.; Hiteshue, R. W. *Am. Chem. Soc., Div. Fuel Chem., Prepr.* (Chicago, Sept., 1970). *15,* A23.
20. Zielke, C. W.; Rosenhoover, W. A.; Gorin, E. In "Shale Oil, Tar Sands and Related Fuel Sources," *Adv. Chem. Ser.* **1976,** *151,* 153.
21. Wood, R. E.; Wiser, W. H. *I & EC Process Des. Dev.* **1976,** *15,* 144.
22. Low, J. Y.; Ross, D. S. In "Organic Chemistry of Coal," *ACS Symp. Ser.* **1978,** *71,* 204.
23. Skinn, J. H.; Vermeulen, T. *Am. Chem. Soc., Div. Fuel Chem., Prepr.* (Honolulu, Apr., 1979) *24*(2), 80.
24. Bugle, R. C.; Wilson, K.; Olsen, G.; Wade, L. G.; Osteryoung, R. A. *Nature* **1978,** *274,* 578.
25. Schlosberg, R. H.; Maa, P. S.; Neavel, R. C. U.S. Patent 4 092 235.
26. Schlosberg, R. H.; Neavel, R. C.; Maa, P. S.; Gorbaty, M. L. *Fuel* **1980,** *59,* 45.
27. Larsen, J. W.; Kuemmerle, E. W. *Fuel* **1976,** *55,* 162.
28. Sawatzky, H.; George, A. E.; Smiley, G.T.; Montgomery, D. S. *Fuel* **1976,** *55,* 16.
29. Schmerling, L.; Ipatieff, V. A. U.S. Patent 2 3888 937.
30. Mobley, D. P.; Salin, S.; Tanner, K. J.; Taylor, N. D.; Bell, A. T. *Am. Chem. Soc., Div. Fuel. Chem., Prepr.*, (Miami Beach, Sept. 1978) *23*(4), 138.
31. Ignasiak, T.; Samman, N.; Bimer, J.; Montgomery, D. S.; Strausz, O. P. submitted for publication in *Fuel.*
32. Mobley, D. P.; Bell, A. T. *Fuel* **1979,** *58,* 661.
33. Ignasiak, B.; Carson, D.; Gawlak, M. *Fuel* **1979,** *58,* 833.
34. Bodily, D. M.; Lee, S. H. O.; Wiser, W. H. *Am. Chem. Soc., Div. Fuel Chem., Prepr.* (Los Angeles, Mar.–Apr., 1974) *19,* 163.

RECEIVED June 23, 1980.

12

Programmed Pyrolysis, Programmed Combustion, and Specific Nitrogen and Sulfur Detection

A New Tool for Studying Heavy Ends of Petroleum and Bitumens

P. VERCIER

Total-Compagnie Francaise de Raffinage, Total Technique, Research Center, Boîte Postale 27, 76700 Harfleur France

When submitting a sample of petroleum distillation residue to a programmed pyrolysis, some nitrogen and sulfur is detected by specific detectors in the vapor evolved ("Type I Nitrogen" or "Type I Sulfur"). The carbonaceous matter remaining at the end of the first step still contains nitrogen and sulfur ("Type II Nitrogen" or "Type II Sulfur") as demonstrated by programmed combustion. Nitrogen forms more Type II structures than does sulfur. This is also true for fractions obtained through precipitation of nonsolvents (asphaltenes, resins, and deasphaltened oils). Type II elements are more selectively concentrated in asphaltenes than in any other fraction, and their content shows a good correlation with asphaltene content. It is believed that Type I elements are attached to molecules of a low molecular weight that distill during heating or to molecular fragments broken by heat treatment, while Type II elements belong to those molecules that tend to form coke in the first step.

Intensive research on elemental analysis has been carried out in this laboratory to overcome difficulties in determining nitrogen and sulfur in heavy fractions of petroleum or bitumens. With existing combustion techniques it was nearly impossible to obtain good overall balances for these elements when individual fractions coming from liquid chromatography were analyzed. Recoveries of 85%–95% of the quantity originally present in the

0065-2393/81/0195-0203$05.00/0

feedstock were obtained when summing up the contents of some forty or more different subfractions (*1*). It was assumed that the combustion was somewhat incomplete. Another explanation for these incorrect results could be an overloading of the sensor. When large amounts of gases are generated suddenly during combustion, instantaneous concentration of the species to be measured can go beyond the capability of the system, either because of slow detector response time or because of the concentration going into the nonlinear range of the device.

To solve the first difficulty, it was thought that more drastic conditions would be required to obtain complete transformation of these organically bound elements. A sample weight larger than the usual milligram size would diminish cumulative errors. But at the same time, this approach would increase the second difficulty by favoring flash combustion with subsequent loss of gases and/or detector overloading.

The method of Oita (*2*) on semi-micro samples for carbon and hydrogen determination, derived from the method of Glass and Cowell (*3*) on macro samples, which uses temperature programming of the sample under inert gas (pyrolysis) followed by oxidation of the resulting vapors in a stream of oxygen, was the base of our research for the determination of nitrogen and sulfur. Ma (*4*) has already pointed out the importance of controlling the conditions of combustion and especially recommends heating the sample in two steps for nitrogen determinations. With this step wise technique, distillation occurs in the sample, the lighter portions leaving first, and combustion occurs smoothly without strain on the sensor.

During tests made with the apparatus built according to this principle, it was observed that in heavy residues, or in asphaltenes and resins, a certain amount of NO (or SO_2) was evolved, but that a solid, charry residue still remained in the combustion boat. To remove all organic matter it was necessary to re-run a temperature programming, this time under oxygen. More NO (or SO_2) was detected during this period. This confirms results of previous observations (*5–9*).

The portion of the element appearing during the first step was called "of Type I," while the remaining part released during combustion was called "of Type II". These preliminary results for nitrogen were presented in Lahnstein, Germany, in October 1979 (*10*). More results are given in (*11*).

This chapter describes the status of the technique and the results obtained so far. Some assumptions will be made to try to explain the postulated existence of two main types of nitrogen- or sulfur-containing species in heavy ends of petroleum and bitumens.

Experimental

Apparatus for Determining Types of Nitrogen and Sulfur. One apparatus is needed for each element under investigation. Each consists of two parts: a mineralization section and a detection section. The first part is the same for nitrogen

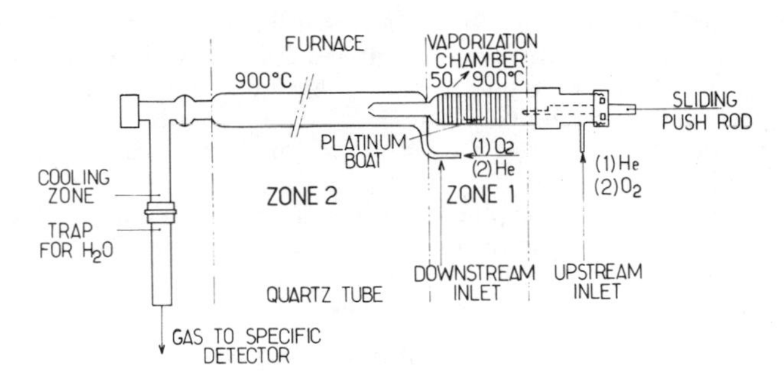

Figure 1. Quartz tube for mineralization

and sulfur devices, but detection is specific. It was not possible to use a single apparatus for both determinations because different flow rates and gases were needed for detection.

MINERALIZATION SECTION. The mineralization section is illustrated in Figure 1. A detailed description will be given elsewhere (*12*). It consists of a quartz tube fitted with two gas inlets: one upstream, ahead of the vaporization chamber (Zone 1), another downstream at the entrance of the large portion of the tube (Zone 2). A large Kanthal wire (Cr + Al + Co + Fe, 3.75 Ω/m) is coiled around the vaporization chamber and a furnace is placed around the main portion of the tube. The tube is closed by a screw cap fitted with a sliding push rod for moving the sample boat inside the programmed temperature section of the tube (the liquid version of this apparatus uses a septum through which a syringe needle can be pushed, the vaporization chamber temperature being kept constant during the whole assay).

During the first step helium (or argon) is swept through the upstream inlet and oxygen through the other inlet, while the vaporization chamber is at room temperature and the furnace at 900° C. The sample is pushed into the vaporization chamber and a voltage is applied to the wire. An electronic temperature programmer using switching transistors receives data from a platinum resistance thermometer positioned between the tube and the heating wire. It compares with the program and feeds power to the system to match the desired temperature. Total time and final temperature are fixed and reproducible from run to run. Three heating rates are available: 50°, 140°, and 400°/min. In this work, speed number two was used. Total time for reaching 900° C: 6 min, 28 s ± 1 s.

Vapors from the more volatile parts or pyrolyzed portions of the sample are flushed past the chamber by the helium (or argon) flow into the jet in Zone 2 where oxygen is present and combustion occurs. As soon as the temperature reaches 900° C, heating the vaporization chamber is discontinued and the sample is cooled at room temperature with a small electrical fan.

When a temperature of 50° C or less is obtained, the second step starts by first switching the gases: oxygen is admitted in the upstream inlet, while helium (or argon) flushes the large tube of Zone 2. Another programmed heating is then activated to 900° C. During this heating period, the tar still present after the first step is completely burned out and more NO (SO_2) is produced.

DETECTION SECTION

Detection of NO. Nitrogen monoxide, dried on a magnesium perchlorate trap, is detected after reaction with ozone by the chemiluminescence produced by excited NO_2 molecules coming back to ground state. Parks and Marietta (*13*) and Drushell (*14*) were the first to use this method for nitrogen determination. The detector used in this study, Analyseur AC 3, was obtained from Environnement S.A., Poissy, France (the commercial apparatus for performing nitrogen analysis—mineralization and NO detection—called Lumazote, is sold by Eraly, Noisy-le-Roi, France). Since the NO_2 to NO conversion unit was removed to prevent scattered results, a response factor is determined every day by burning a sample of tomatin (Fluka, Germany, nitrogen content: 1.354%).

The counts detected by the photomultiplier are integrated during the run and an analog display is given on a pen recorder. Figure 2 shows four examples of such traces with peaks for Type I and Type II nitrogen.

Total nitrogen content is given by the following formula:

$$\% \mathrm{N_T} = \frac{\text{total number of counts obtained during the run}}{\text{number of counts for 1 ng nitrogen} \times 10^4 \times \text{sample weight, in mg}}$$

Type II nitrogen content can be expressed either as the ratio of Type II peak area (in integrator counts) to the total area of all peaks, multiplied by 100, which is a relative value, or as an absolute value obtained by multiplying total nitrogen content by the relative value of Type II nitrogen.

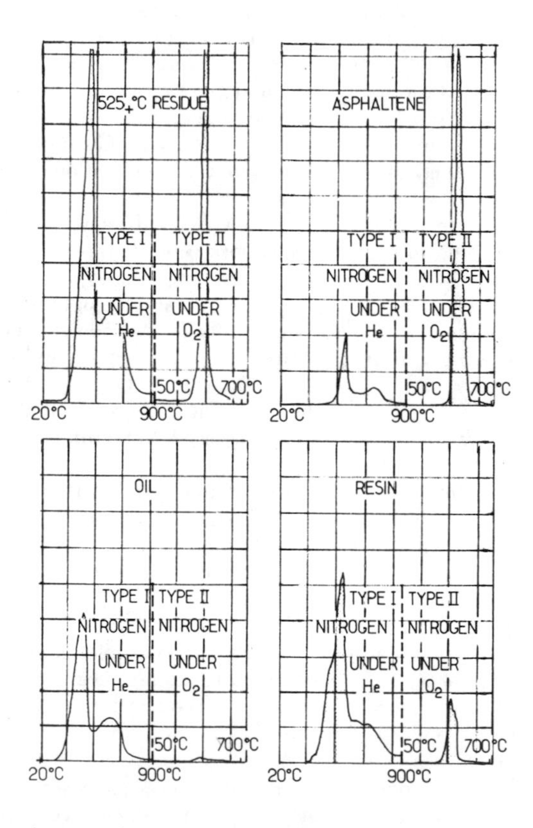

Figure 2. Distribution of Type I and Type II nitrogen in fractions of Oficina crude oil

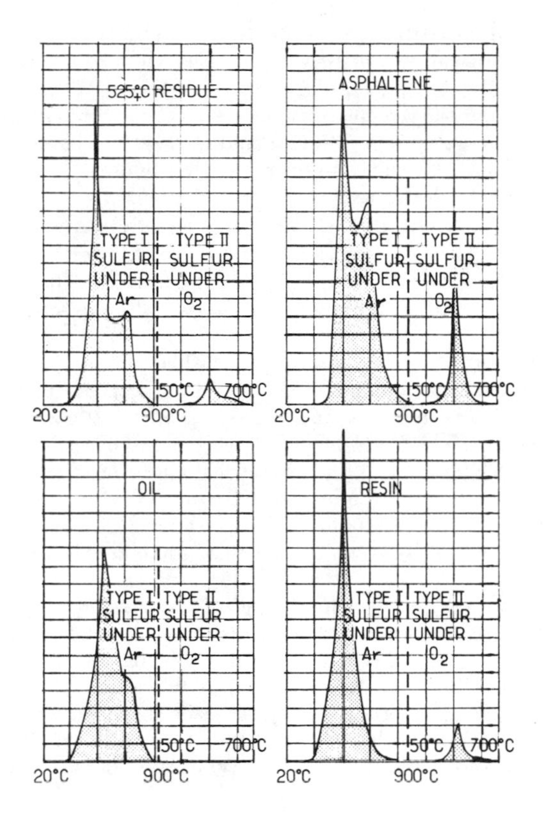

Figure 3. Distribution of Type I and Type II sulfur in fractions of Oficina crude oil

Type I nitrogen values are obtained in the same way.

The measurement range for solids and/or viscous liquids is from 10 ppm to more than 10% without dilution.

Detection of SO_2. SO_2 is detected by a photomultiplier tube that collects fluorescence in the 3500-Å range when the gas is excited by UV radiation at 210 nm (*15*). The detector used in this study, Analyseur AF 20, was obtained from Environnement S.A., Poissy, France. To take into account SO_3 formation, not detected by this system, a response factor is determined every day by burning a sample of Bright Stock Solvent, the sulfur content of which has been measured a number of times by Wickbold combustion, combustion and coulometry (*16*) and x-ray fluorescence. The counts are integrated as a function of time during the two heating periods. Examples of traces for sulfur are given in Figure 3. Calculations are done in the same way as for NO.

Results will be given in total sulfur content, relative and absolute Type II (or Type I) sulfur.

The measurement range for this apparatus on solids and/or viscous liquids is from 50 ppm to at least 8%.

Because the sulfur detector is more sensitive than the nitrogen detector and the sulfur content of samples higher, the sample weight is often under 1 mg compared with 4–5 mg in the case of nitrogen.

Preparation of Distillates and Residues. Crude oil samples were topped up to 350° C in an isothermal distillation unit (*17*). The 350+ ° C residues were treated in a short path distillation unit (*18*) to prepare 300°–400° C and 400°–525° C or 400°–550° C distillates. The boiling range was checked by simulated distillation using a technique developed in common with ELF and I.F.P. (*19*). The residues (525+ ° C or 550+ ° C) used in this work come from the same operations.

Preparation of Asphaltenes

HEPTANE ASPHALTENES (C_7). Asphaltenes were prepared according to References *20, 21* in boiling heptane with a ratio of one part by weight of residue to thirty parts by volume of solvent (for example 1 g/30 mL). In the United States there is a proposed method for asphalt composition analysis (*22*) using *n*-heptane and a solvent/charge of 100 (v):1 (w).

PENTANE ASPHALTENES (C_5). No established method is available for the preparation of pentane asphaltenes. ASTM D-2006 was discontinued in 1976. Another ASTM method (for rubber extender and processing oils) (*23*) uses a 10-g sample and only 100 mL of pentane, which is insufficient for a correct dispersion of the sample into the solvent. A very precise method for these asphaltenes from coal-derived liquids has been described (*24*), but some might object to the use of benzene because of the solubilizing properties of this solvent towards asphaltenes.

Our method is derived from the standardized method used in Europe, replacing *n*-heptane by *n*-pentane.

Preparation of Resins. Maltenes (1 g) obtained after removal of asphaltenes (obtained either by C_5 or C_7) and solvent were dissolved in cyclohexane and the solution was run on a dual column (Davison Silicagel grade 62 and Alcoa F 20 alumina, according to USBM–API method)(*25*). After eluting the saturate and aromatic fractions ("deasphalted oil") with appropriate solvent, resins on the column were extracted once with a 50/50 (v/v) mixture of ether and methanol, then with a 75/25 (v/v) mixture of chloroform and methanol, last with a 75/25 (v/v) mixture of carbon tetrachloride and methanol (*26, 27*).

This method was preferred to a previous one using Attapulgus clay in batch (*28*) because it was faster.

Samples were considered for further work only when the material balance of saturates, aromatics, and resins was within 1% of total recovery.

Weight Losses. Weight losses were determined on a Perkin-Elmer model TGS-2 thermobalance, using either nitrogen or air as sweeping gas.

Results and Discussion

Factors Influencing Results. Pyrolysis by progressive application of heat during a limited period of time under an inert atmosphere brings about a partial destruction of the sample. The choice of time and final temperature influences the results, and a compromise has to be found to obtain an effect large enough to provide significant changes in the sample but small enough to leave behind material for further oxidation.

The results of the conditions chosen for this work can be seen on the solid curve (He) in Figure 4, which shows weight losses of an asphaltene sample in a thermobalance under nitrogen atmosphere, with a heating rate matching the one used in the nitrogen or sulfur determination. At the end of the heating period (to about 900° C), nearly 25% of the initial sample remains in the boat. Ritchie et al. (*29*) show a similar curve for Athabasca asphaltenes. The weight loss at 700° C as calculated from their curve amounts to 40% of the starting material.

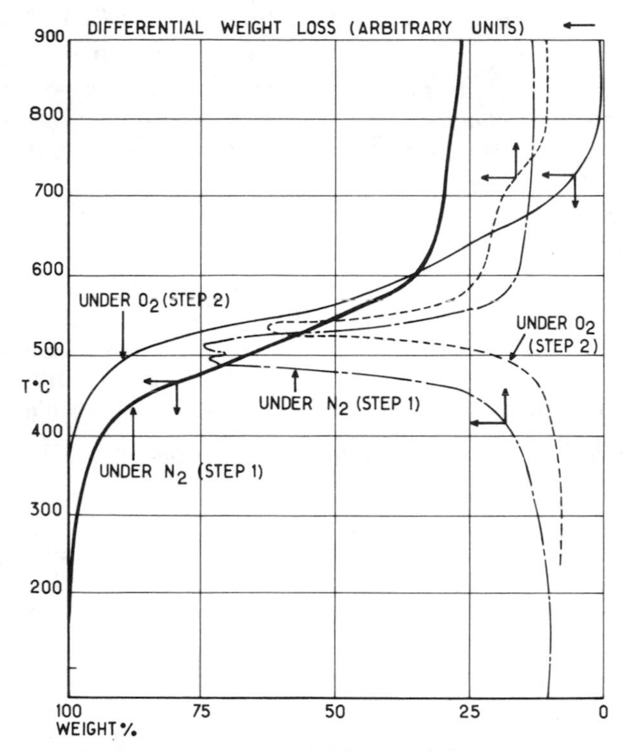

Figure 4. Thermogravimetric and differential thermogravimetric curves of the asphaltenes (C_5) from a 525+° C Oficina residue

This is a confirmation of the findings of these authors showing an early decomposition for asphaltene beginning slightly above 200° C and reaching a maximum rate at 500° C.

But close examination of the differential thermogravimetric trace reveals that, although major losses occur at 500° C, some product disappears slowly between 600° and 900° C. This might be attributable to the beginning of a breakdown of the more fragile parts of the molecules (*29*). This is the reason why it is so important to closely control time and temperature in the program, or at least to hold them constant from run to run.

Even so, repeatability is not as good for Type II nitrogen or Type II sulfur as it is for total nitrogen or for total sulfur. Tables I and II show results obtained on an Oficina 525+° C residue and its fractions issued from *n*-pentane deasphalting (Table I for nitrogen, Table II for sulfur). Overall balance on total nitrogen at the asphaltene and maltene level is 100.3%; at the asphaltene, resin, and oil level it is 98.3%. The respective values for sulfur are 100% and 99.8%. When making the same calculations for Type II elements, recoveries are 114% and 110% for nitrogen II and 95% and 117% for sulfur II. Other values are given in the notes of Table III for nitrogen on an Aramco 525+° C residue and fractions obtained by *n*-C_7 and *n*-C_5 deasphalting.

Table I. Nitrogen Measurement—Oficina Residue (525+ ° C)[a]

Residue or Fraction	*Mass Balance (wt %) (1)*	*Total Nitrogen (wt %) (2)*	*Type II Nitrogen (Relative) (%) (3)*	*Type II Nitrogen (Absolute) (wt %) (4)*
Residue	100	0.705	31.8	0.224
Asphaltenes (C_5)	20.9	1.366	73.9	1.01
Maltenes (C_5)	79.3	0.532	10.3	0.055
Resins (C_5)	22.6	1.137	11.5	0.131
Oils (C_5)	56.0	0.269	4.7	0.0126

Total Nitrogen Balance		*Distribution of Total Nitrogen in Fractions[d]*	
At the maltene level	100.3%	Asphaltenes	41.1
At the resin and oil level	98.3%[b]	Resins	37.1
		Oils	21.8

Type II Nitrogen Balance		*Distribution of Type II Nitrogen in Fractions[e]*	
At the maltene level	114 %	Asphaltenes	85.2
At the resin and oil level	110.5%[c]	Resins	11.9
		Oils	2.8

[a] 35.3% weight of crude oil.

[b] $$\Sigma_{N_T} = \left[\frac{(1) \times (2)}{100}\right]_{asph} + \left[\frac{(1) \times (2)}{100}\right]_{resins} + \left[\frac{(1) \times (2)}{100}\right]_{oils}.$$

[c] $$\Sigma_{N_{II}} = \left[\frac{(1) \times (4)}{100}\right]_{asph} + \left[\frac{(1) \times (4)}{100}\right]_{resins} + \left[\frac{(1) \times (4)}{100}\right]_{oils}.$$

[d] 100% = Σ_{N_T}, not the measured value on residue, i.e., 0.693, not 0.705.
[e] 100% = $\Sigma_{N_{II}}$, not the measured value on residue, i.e., 0.248, not 0.224.

This clearly demonstrates that Type II determinations are less precise than their absolute counterparts and that real values should be within ± 10% of the quoted values for nitrogen (± 20% for sulfur). As an example, relative Type II nitrogen content in Aramco residue was found to be 24.8% (Table III) while in another experiment the value was 19% (Table IV, obtained during the first assays. Compare with the asphaltenes content in the same table: 5.0% and 5.3%).

A possible explanation for these variations in the case of sulfur could be weighing errors: sample weights are around 0.5 mg, attributable to the very high sensitivity of the sulfur detector. This can be corrected by adding supplementary attenuation in the signal response. For this same reason, one cannot detect Type II sulfur in maltenes, although it is found in resins after further fractionation. A possible way of overcoming this difficulty is to make two successive determinations, one for measuring total sulfur (Types I and II) and another, on a large sample, for specifically measuring sulfur II.

A small variation in the heating period should not be of importance, even with the fast rates used (± 1 s brings only a 5° C difference in final temperature). A more general explanation is mass effects, although variations from 1 to 5 mg in sample weight (for nitrogen) failed to indicate significant and parallel changes in Type II nitrogen content (from 29.5 to 32.2). The ratio volume to surface of sample could also be of importance: this point needs clarification in the future.

With these confidence limits in mind, let us examine now where these types of structures are to be found and their respective behavior.

Table II. Sulfur Measurements—Oficina Residue (525+ ° C)[a]

Residue or Fraction	*Mass Balance (wt %) (1)*	*Total Sulfur (wt %) (2)*	*Type II Sulfur (Relative) (%) (3)*	*Type II Sulfur (Absolute) (wt %) (4)*
Residue	100	2.85	7.0	0.200
Asphaltenes (C_5)	20.9	3.82	23.1	0.882
Maltenes (C_5)	79.3	2.58	0.4	0.01
Resins (C_5)	22.6	3.11	7.3	0.228
Oils (C_5)	56.0	2.40	0.0	0.0

Total Sulfur Balance		*Distribution of Total Sulfur in Fractions[d]*	
At the maltene level	100.0%	Asphaltenes	28.1
At the resin and oil level	99.0%[b]	Resins	24.7
		Oils	47.2

Type II Sulfur Balance		*Distribution of Type II Sulfur in Fractions[e]*	
At the maltene level	95.5%	Asphaltenes	78.3
At the resin and oil level	117.5%[c]	Resins	21.7
		Oils	0

[a] 35.3% weight of crude oil.

[b] $$\Sigma_{S_T} = \left[\frac{(1) \times (2)}{100}\right]_{asph} + \left[\frac{(1) \times (2)}{100}\right]_{resins} + \left[\frac{(1) \times (2)}{100}\right]_{oils}.$$

[c] $$\Sigma_{S_{II}} = \left[\frac{(1) \times (4)}{100}\right]_{asph} + \left[\frac{(1) \times (4)}{100}\right]_{resins} + \left[\frac{(1) \times (4)}{100}\right]_{oils}.$$

[d] 100% = Σ_{S_T}, not the measured value on residue.

[e] 100% = $\Sigma_{S_{II}}$, not the measured value on residue.

Table III. Nitrogen Measurement—Aramco Residue (525+ ° C)[a]

Residue or Fraction		*Mass Balance (wt %)*		*Total Nitrogen (wt %)*	*Type II Nitrogen (Relative) (%)*	*Type II Nitrogen (Absolute) (wt %)*
Residue		100		0.350	24.8	0.0868
Asphaltenes	nC_7	5.0	99.2	1.017	79.0	0.803
Resins	nC_7	19.5		0.948	23.7	0.225
Oils	nC_7	74.7		0.134	4.8	0.006
Asphaltenes	nC_5	9.9	100.2	0.914	71.7	0.655
Resins	nC_5	18.3		0.902	16.1	0.145
Oils	nC_5	72.0		0.108	4.6	0.005

Total Nitrogen Balance[b]

nC_7	96.7%
nC_5	95%

Distribution of Total Nitrogen in Fractions[b]

	nC_7	nC_5
Asphaltenes	15.8	27.0
Resins	54.6	49.5
Oils	29.6	23.4

Type II Nitrogen Balance[b]

nC_7	102.3%
nC_5	109%

[a]23.1% weight of crude oil.
[b]As defined in Table I.

Occurrence of Type II Nitrogen and Sulfur. The bulk of information on these two types of elements has been obtained on nitrogen; therefore, emphasis will be placed on this species.

It was first shown that no Type II nitrogen exists in any of the 400°–525° C or 400°–550° C distillates from the oils investigated so far.

For the residues, Type II nitrogen is universally present in crudes from different parts of the world, as shown in Table IV. Their absolute quantity is not very large, but relatively speaking it goes from a low 5.1% of the total nitrogen content in the Forcados residue to a high 36.1% in the Souedie residue (although initial boiling point in the residue is 550° C instead of 525° C). These values have been arranged by increasing Type II content. There is no direct correlation with total nitrogen content, but a relationship has been observed between Type II nitrogen (absolute) and C_7 asphaltene content. The correlation coefficient is 0.86, but when deleting the Oficina outlier, the fit is better ($r = 0.96$). This same factor is obtained when correlating C_7 asphaltene content and relative Type II nitrogen without suppression of any data, while there is no apparent correlation with vanadium or nickel content.

Table IV. Total Nitrogen, Type II Nitrogen, Vanadium and Nickel Contents in Several Residues

Residue from Crude Oil	*Total Nitrogen (%)*	*Type II Nitrogen (%, Relative)*	*Type II Nitrogen (%, Absolute)*	*Asphaltenes C_7 (wt %)*	*Vanadium (ppm)*	Nickel *(ppm)*
Forcados 525+° C (Nigeria)	0.764	5.1	0.048	0.13	10	52
Oural 525+° C (USSR)	0.592	11.3	0.067	3.6	166	55
Aramco 525+° C (Saudi Arabia)	0.350	19	0.066	5.3	69	18
Iraq 550+° C (Iraq)	0.444	20.5	0.090	8.5	142	65
Handil 525+° C (Indonesia)	0.370	27.2	0.100	12.6	7	23
Oficina 525+° C (Venezuela)	0.712	29.5	0.219	13.0	360	89
Basrah 525+° C (Iraq)	0.502	31.0	0.155	16.8	253	79
Souedie 550+° C (Syria)	0.529	36.1	0.203	21.7	270	91

Respective Behavior of Nitrogen- and Sulfur-Bearing Structures Under Programmed Pyrolysis. Amounts of Type II element can be seen as a measure of the difficulty for breaking nitrogen–carbon and sulfur–carbon bonds. It is assumed that part of the residue is still volatile and, therefore, distills when the temperature is high enough (*29*, *30*), forming vapors removed from the sample boat and subsequently burned in the oxidation section. Simultaneously, some cracking could occur in the bulk of the sample (*30*)—with possible side reactions—yielding fragments with heteroelements still bound to them, which are easily swept off into the next section of the apparatus. The other part of the sulfur or nitrogen, which is bound to more stable basic units, cannot break loose because of coke formation through dehydrogenation and ring condensation (such a mechanism has been postulated by Yen (*31*)). Only the most severe conditions—high temperatures, high oxygen content—can bring final destruction of these highly resistant species (*32*).

It can be seen from the results given in Tables I and II and illustrated in Figure 5 that nitrogen and sulfur behave in quite different ways. The percentage of original sulfur remaining as Type II compounds is much smaller than the percentage of original nitrogen remaining as Type II nitrogen. This is true for the residue and for the various fractions obtained by treatment with *n*-pentane. Only 7% of the total sulfur is held back in the tar at the end of the

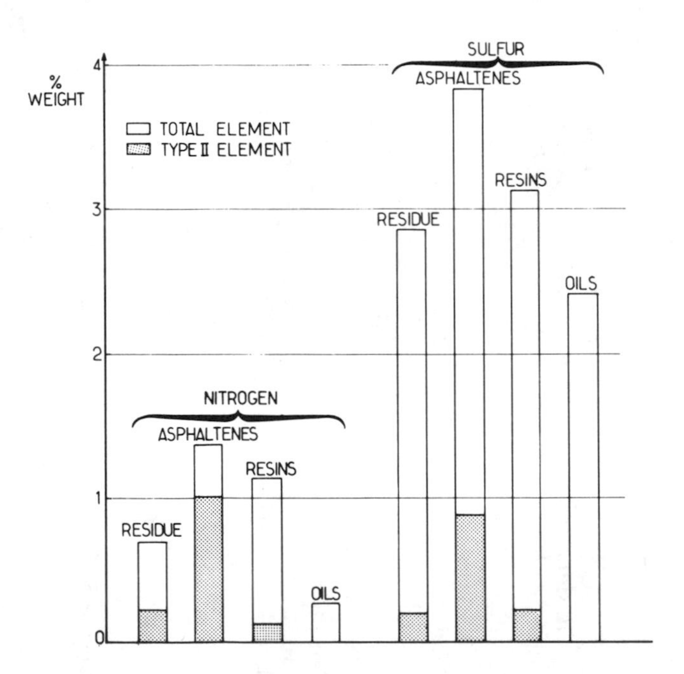

Figure 5. Respective amounts of Type II vs. total content of elements: Oficina residue (525+° C)

first step while 31.8% of the total nitrogen remains in the coke after the same treatment. Even more striking are the figures in asphaltene fractions: 23.1% of the sulfur compared with 73.9% of the nitrogen remains after the first step.

The enrichment in sulfur in the vaporized portion of the sample (*29*) and in nitrogen in the coke (*33*), and correlatively the low content of nitrogen in the volatiles (*29*) has already been noted, although it was not realized through lack of precise data to what extent the phenomenon occurred especially for the relative depletion of sulfur in the coke.

Sulfur molecules in residues are more readily destroyed and less prone to form resistant species than are nitrogen-containing structures. The presence of sulfur in volatiles has been interpreted by Ritchie et al (*29*) as evidence of thiophenic forms in the original asphaltenes. Further work, which is under way, is necessary to find out in what structures nitrogen is bound in the residue and what type of reaction occurs during pyrolysis.

Concentration of Type II Nitrogen or Type II Sulfur in Subfractions. When separating distillation residues of Oficina crude oil into asphaltenes, resins, and deasphalted oils using *n*-pentane, some drastic changes occur in nitrogen and sulfur distribution, as can be seen in Tables I and II.

For total nitrogen, 41.1% of the element is present in the asphaltene portion of the residue, 37.1% in the resins, and 21.8% in the oil. For Type II nitrogen (which in the feed represented 31.8% of total nitrogen content), the respective values are 73.9%, 11.5% and 4.7%. In other words, 85.2% of the total quantity of Type II nitrogen present in the residue is concentrated in the asphaltene portion while only 11.9% remains in the resins and a mere 2.8% is left in the oil.

Table V further shows that vanadium and nickel distributions in fractions closely follow the pattern of Type II nitrogen or Type II sulfur: 82%, 17.4%, and 0.6% of the vanadium originally present in the residue are found later in, respectively, asphaltenes, resins, and oils (obtained through C_5 precipitation).

This suggests the existence in residues of molecules having vanadium and nickel bound in a constant ratio to Type II nitrogen and Type II sulfur atoms that could be precursors of coke formation (*34*). These molecules or structures are more or less selectively precipitated by *n*-pentane into the asphaltenes, although a certain quantity still remains in the resins.

It would be useful to pursue the investigations by building two other devices with specific carbon and oxygen detection to try to get a rough idea of the elemental composition of these types of molecules.

When using different *n*-alkanes for precipitation, distribution of total nitrogen is about the same in all cases (*see* Table III): nearly half of the total nitrogen is concentrated in the resin fraction while Type II concentration in asphaltenes increases when going from C_5 to C_7. This indicates that compounds with Type II elements are more abundant in higher molecular weight species and that C_5 precipitation dilutes them with other types of molecules.

Table V. Distribution of Heteroelements in Oficina Fractions (Wt %)

Element	*Total Nitrogen*	*Type II Nitrogen*	*Total Sulfur*	*Type II Sulfur*	*Vanadium*	*Nickel*
Asphaltenes (C_5)	41.1	85.2	28.1	78.3	82.0	79.0
Resins (C_5)	37.1	11.9	24.7	21.7	17.4	15.1
Oils	21.8	2.8	47.2	0	0.6	5.9
Residue	100	100[a]	100	100[b]	100	100

[a] 68.2% Type I and 31.8% Type II nitrogen in the feed.
[b] 93.0% Type I and 7.0% Type II sulfur in the feed.

Conclusion

The complexity of heavy fractions of crude oils suggests that there are many types of nitrogen and sulfur compounds present. The wide range of existing types has been divided into two groups (Type I and Type II) by the described analytical procedure.

That part of the element called "of Type I" is believed to be attached to the more volatile portion of the residue or to fragments produced during pyrolysis from less stable molecules, while the other part ("of Type II") cannot escape from the coke being formed during the first step because it is more and more deeply buried in condensating structures. However, further work is necessary to identify these nitrogen- or sulfur-bearing molecules or fragments, and even more for studying the tar-like material.

This programmed pyrolysis and programmed combustion provides a useful tool for studying petroleum residues: measurement and evaluation of these heat resistant structures in various oils from a geochemical point of view; evolution and respective behavior of nitrogen- and sulfur-bearing structures to coke formation on catalysts during hydrotreatment, and estimation of the ease of desulfurization or denitrification, for the process engineer.

Acknowledgments

Appreciation is expressed to the members of Total Research Center, Analytical Department: B. Thiault, D. Joly, R. Antolini, M. Mouton, J.M. Colin, A. Allain for numerous helpful discussions; to R. Chelveder for designing and S. Varnier for building the prototypes for nitrogen and sulfur determinations; to J. Goupil for preparing the various distillates and residues; to S. Contremoulins for separating asphaltenes, resins, and oils; to D. Leborgne and P. Duvauchelle for running, respectively, the nitrogen and sulfur measurements.

I also want to extend my thanks to P. Goursot (CNRS, Centre de Recherches Physico-Chimiques de la Surface du Solide, Mulhouse, France), who performed the thermogravimetric determinations and suggested many useful directions for research. I express my gratitude to the Management of Total Research for granting permission to publish this paper, especially to its Scientific Director, J. Grosmangin, for his continuous interest in the work.

Literature Cited

1. Nambi Aiyar, V.; Houwen, O. H.; Bachelor, F. W. *Fuel* **1980,** *59,* 276.
2. Oita, I. J., Standard Oil Company (Indiana), Naperville, Ill., personal communication.
3. Glass, J. R.; Cowell, W. H. *Anal. Chem.* **1971,** *43,* 678.
4. Ma, T. S. "Modern Organic Elemental Analysis"; Dekker: New York, 1979; p. 113.
5. Yen, T. F. *Am. Chem. Soc., Div. Fuel Chem., Prepr.* (Los Angeles, Mar.–Apr., 1971), *15*(1), 57.
6. Clerc, R. J.; O'Neal, M. J. *Fuel* **1970,** *49,* 76.
7. Moschopedis S. E.; Parkash, S.; Speight J. G. *Fuel* **1978,** 57, 431
8. Speight, J. G.; Penzes, S. *Chem. Ind.* **1978,** 729.
9. Speight, J. G.; Moschopedis, S. E. *Am. Chem. Soc., Div. of Pet. Chem., Prepr.* (Wash., D.C., Sept., 1979) *24*(4), 919.
10. Dolle, J. M.; Leborgne, D.; Vercier, P. *Erdöl Kohle* **1980,** *33*(4), 175.
11. Allain, A.; Vercier, P.; Leborgne, D. *Journées de Calorimétrie et d'Analyse Thermique, Prepr.* (Barcelona, Spain, June, 1980).
12. Dolle, J. M.; Leborgne, D.; Chelveder, R.; Varnier, S., unpublished data.
13. Parks, R. E.; Marietta, R. L. US Patent 4018562, 1977.
14. Drushel, H. V. *Anal. Chem.* **1977,** *49,* 932.
15. Syty, A. *Anal. Chem.* **1973,** *45,* 1744.
16. Solea-Tacussel, Villeurbanne, (Coulomax apparatus) French Patent 2227794, 1973.
17. Mouton, M.; Robinet, A.; Madec, M. *Rev. Inst. Fr. Pétrole* **1978,** *33*(2), 299.
18. Vercier, P.; Mouton, M.; *D. G. M. Koblentz* **80,** 211; *Analusis,* in press.
19. Petroff, N.; Colin, J. M.; Feillens, N.; Follain, G. Rev. Inst. Fr. Petrole, in press.
20. French standard, AFNOR NF-T-60115.
21. British standard, IP 143/77.
22. ASTM Committee D-4, 1978, *15,* 1157.
23. ASTM Committee D-2, D 2007-75.
24. Mima, M. J.; Schultz, H.; McKinstry, W. E., PERC/RI 76/6, 1976, Pittsburg Energy Research Center, Pittsburgh, PA.
25. Jewell, D. M.; Weber, J. H.; Bunger, J. W.; Plancher, H.; Latham, D. R. *Anal. Chem.* **1972,** *44,* 1391.
26. Vercier, P.; Colin, J. M. *CNRS, Colloq. Produits Lourds Pétrole, Prepr.* (Marseille, France, Oct., 1979).
27. Vercier, P.; Colin, J. M.; *Rev. Inst. Fr. Pétrole* **1981,** *36*(1), 45.
28. Vercier, P.; Thiault, B.; Colin, J. M.; Mersseman, M., *DGMK Compendium 77/78, Vorträge der 26. DGMK Haupttagung;* (Berlin, Oct., 1978), 1499.
29. Ritchie, R. G. S.; Roche, R. S.; Steedman, W. *Fuel* **1979,** *58,* 523.
30. Ritchie, R. G. S.; Roche, R. S.; Steedman, W. *Ind. Eng. Chem.-Prod. Res. Dev.* **1978,** *17*(4), 370.
31. Yen, T. F., Ed. "The Role of Trace Metals in Petroleum"; Ann Arbor Science: Ann Arbor, 1975; p. 22.
32. Starr, J., personal communication.
33. Speight, J. G. *Fuel* **1970,** *49,* 134,
34. Furimsky, E. *Ind. Eng. Chem.-Prod. Res. Dev.* **1978,** *17*(4), 329

RECEIVED June 23, 1980.

Characteristics of Tar Sand Bitumen Asphaltenes as Studied by Conversion of Bitumen by Hydropyrolysis

JAMES W. BUNGER and DONALD E. COGSWELL

Department of Mining and Fuels Engineering, University of Utah, Salt Lake City, UT 84112

Chemical characteristics of tar sand bitumen asphaltenes were examined by inference from spectroscopic and hydropyrolysis processing studies of virgin bitumen and deasphaltened bitumen. While average properties differ between asphaltenes and maltenes, major overlap between these fractions is experienced with respect to carbon type, heteroatom type, molecular weight, and volatility. Theoretical considerations suggest that given species are partitioned between the maltene and asphaltene portions further contributing to a lack of chemical distinction between maltene and asphaltene species. When subjected to conversion by hydropyrolysis, both maltenes and virgin bitumen yielded products exhibiting pentane insolubles with maltenes yielding about 11% and asphaltenes yielding (by calculation) 20% insolubles. A strong precursor–product relationship between asphaltenes and pentane insolubles in the product does not exist. No chemical features (e.g., carbon type, heteroatom type, molecular weight, or structure) were found that distinguished species found in the asphaltenes fraction from those in the maltenes fraction. Thus, asphaltenes appear to be chemically nondistinct from maltenes and correlations between asphaltene content and processability, while useful to an engineer, are chemically fortuitous.

Asphaltenes have been the subject of considerable discussion and controversy in the literature. Controversy and ambiguity arise largely because of the lack of chemical definition of asphaltene mixtures for which composition is dependent upon the source material and method of isolation. While

0065-2393/81/0195-0219$0.500/0

empirical correlations often exist between increased asphaltene content and increased difficulty in processing of feedstocks, the chemical basis of these correlations remains to be demonstrated.

It is possible that a cause-and-effect relationship between molecules comprising the asphaltene fraction and difficulty in processing does not exist and that these relationships are largely fortuitous in that the chemistry involved in asphaltenes isolation is only secondarily related to chemistry of processing. If such is the case, much of the ambiguity surrounding asphaltenes derives from the incorrect assumptions about the chemistry of complex hydrocarbon mixtures inferred from empirical correlations.

In this work, we approached the problem of chemistry and definition of asphaltenes with the following hypothesis: (i) asphaltenes are not structurally unique from nonasphaltenes; (ii) average molecular weights for asphaltenes are not appreciably higher than those for nonasphaltenes, and strong associative forces account for the inordinately high measured values for asphaltene molecular weights; and (iii) the in-situ properties of those molecules comprising the asphaltenes fraction differ considerably from those manifest when isolated in an asphaltene fraction.

To test these hypotheses, a tar sand bitumen containing 20 wt % pentane asphaltenes was characterized and processed by hydropyrolysis before and after removal of asphaltenes. Product yields and structure were determined and the influence of asphaltenes on results was determined by inferrence. Feedstocks and products were characterized according to elemental analysis, physical properties, simulated distillation, and carbon-type analysis. Inferences made in this study are discussed in the context of the reported literature.

Experimental

Sample Source and Preparation. The Sunnyside tar sand sample was freshly mined from the old asphalt quarry east of Sunnyside, Utah. Bitumen was extracted with benzene, filtered, and the solvent removed as previously described (*1*). Deasphaltening was accomplished using 40:1 (v:w) *n*-pentane-to-sample in which the mixture was stirred and allowed to digest overnight at room temperature. The precipitate was filtered and washed with sparing quantities of chilled *n*-pentane. The solvent was removed from the fractions as before and the respective fractions from three equal-sized deasphaltening runs were combined and used directly for property measurements and as the process feedstock. A third process feedstock was prepared by dissolving a known amount of virgin bitumen in benzene, and combining with this solution 0.7 times the normal amount of asphaltenes. The solvent was removed by rotary evaporation to generate a sample "1.5 normal" in asphaltenes. It was not confirmed that this procedure resulted in completely homogeneous dispersion of asphaltenes in virgin bitumen.

Elemental Analysis and Physical Properties. Elemental analysis was accomplished by conventional microanalytical techniques in a commercial testing laboratory. Density, refractive index, average molecular weight (VPO), Conradson carbon residue, and ash content were determined by standard methods. Viscosity was determined by a cone-plate viscometer. Simulated distillation was accomplished using a ¼″ × 18″ column of Anachrome Q, 3% Dexil 300, programmed from −30 to

+ 350°C at 10°/min and an FID detector, and using as an internal standard an equal volume mixture of C_9–C_{16} *n*-alkylbenzenes. (*See also* References *1* and *2*). Gas analysis was performed on a ⅛″ × 240″ column packed with Chromosorb 102, programmed from −30 to +250°C at 10°/min and an initial isothermal time of 5 min and using a TCD detector.

Carbon-13 NMR Spectroscopy. High resolution ^{13}C NMR spectra of maltenes, bitumen, and liquid products taken up in a 1:2 mixture of $CDCl_3$ were obtained on a Varian XL-100-15 FT spectrometer equipped with a V-4412 probe and a 2.5 megaword disc. A pulse angle of 45° and a repetition period of 60 s were selected to avoid saturation of any nonprotonated carbons. Gated decoupling was also employed to suppress NOE effects in the line intensities. These precautions have been shown for mixtures of model compounds to minimize error in integrated line intensities to less than 10% (*3*).

The combined cross-polarization and magic-angle spinning (CP/MAS) spectra of the asphaltenes were obtained on a single-coil, double-tuned probe similar in design to that of Cross, Hester and Waugh (*4*) but which has been adapted for an electromagnet. Equipped with a D_2O external lock, this probe accepts either 12-mm liquid sample tubes or the magic-angle spinner (*5*). Radiofrequency fields of 17 G and 12 G at 25.16 MHz and 100.06 MHz respectively were obtained at 90 W power. Isolation between the ^{13}C and ^{1}H channels is in excess of 60 db.

The CP/MAS spectra were taken with the single contact (*6*) sequence. The 90° pulse for the ^{1}H spin locking is continuously variable and the amplitude of the ^{13}C irradiation controlled to within 0.1 db for matching the Hartmann–Hahn (*7*) condition.

The high-speed magic-angle spinner used has been described in detail elsewhere (*5*). The particular one we used is constructed of Macor, a machineable glass that gives no background in the ^{13}C CP/MAS spectra. Additional experimental details and a discussion relating this procedure to solid coals and coal derived liquids have been previously published (*8*).

Hydropyrolysis Process. The hydropyrolysis reactor, consisting of a coiled stainless steel tube $^{3}/_{16}$″ i.d. × 236″ long has been previously described (*9, 10*). Average residence time was calculated from the volumetric throughput and reactor volume. Volumetric measurements made at atmospheric pressure and room temperature were corrected to process conditions assuming ideal gas behavior. (The reaction mixture is typically greater than 97 mol % hydrogen, so residence time may be slightly overestimated by this assumption.) Accumulated coke was determined after one or more runs by burning the coke in air and by measuring the CO_2 evolved.

Hydrogen consumption (uptake) was calculated for those runs that produced essentially only liquids and gases by first conducting a material balance on the carbon and calculating hydrogen content from the composition of the carbon-containing gases and the elemental analysis of the liquids. Heteroatoms fed but not found in the organic liquids were assumed to exist in their fully hydrogenated form, that is, NH_3, H_2S, H_2O, for this calculation. Thus, hydrogen consumption is not underestimated by this procedure. For further description of results of hydropyrolysis and calculation of hydrogen consumption, see References *10–12*.

Results

Feedstock Characteristics. The Sunnyside tar sand sample contains 9.3 wt % bitumen. The extracted bitumen was subjected to deasphaltening and results of three runs were 20.6, 19.9, and 21.4 wt % (20.6% average) of the total bitumen. The elemental analysis and physical properties of the bitumen, maltenes, asphaltenes, and the asphaltene-enhanced bitumen are given in Table I. The virgin bitumen is typical of Uinta Basin samples (*14*), which are of extremely high molecular weight and viscosity, but also of high hydrogen content, low sulfur, and appreciable nitrogen content compared with many petroleum residues. The molecules comprising maltenes are of a higher average hydrogen content than those in the virgin bitumen, and the asphaltenes contain a higher average heteroatoms content. The weighted average of the maltenes and the asphaltenes elemental analysis approximates that of the virgin bitumen, lending credence to the analysis results.

Physical properties show the bitumen and subfractions possess high molecular weights and low API gravities. Asphaltene content and carbon residue of the virgin bitumen are somewhat higher than other Uinta Basin bitumens (*14*). The average molecular weight of heavy bitumen fractions, particularly the asphaltenes, is a subject commonly debated (*14–18*). The results given in Table I are not thought to be representative of the true molecular weights, and values as high as 8000 are so erroneous as to be meaningless. The results for the asphaltenes show that the intermolecular associations can be partially disrupted by changing the polarity and solvating properties of the solvent. The apparent higher molecular weight for virgin bitumen exhibited when using pyridine is presently inexplicable. However, it has previously been observed that solvents of higher polarity can lead to a more associated state for nonpolar species (*14*).

The virgin bitumen and fractions were characterized by ^{13}C NMR. Results shown in Figure 1 show that the spectra of the maltenes and the virgin bitumen are virtually superimposable. The high resolution spectra reveal several resolved peaks in the saturates region attributable to *n*-alkyl species, some of which are of notably lower intensity in the maltenes spectrum, but in general, the vast majority of the saturates carbon ($>80\%$) is found under the envelope and is probably due to complex alicyclic and/or isoparaffin structures. These results are consistent with an earlier observation (*1*) that saturated hydrocarbons from Uinta Basin bitumen are high in naphthenes and low in free paraffins.

The ^{13}C NMR spectra of the asphaltenes obtained in the solid state is shown in Figure 2. These results are compared with a spectrum of the maltenes obtained in solution ($CDCl_3$). The solution spectrum of maltenes has been artificially broadened to produce approximately the same resolution as the solid-state spectrum of the asphaltenes. It is very apparent from Figure 2 that aside from the relative intensities of the aromatics vs. the saturates portion that the two spectra are essentially identical. This observation has important implications in that it shows that the gross hydrocarbon structure of asphaltenes (relative abundance of carbon types) is not significantly different from that of the maltenes.

The semiquantitative results of the ^{13}C NMR are given in Table II. These results must be considered preliminary at this time since evidence has not been

Table I. Elemental Analysis and Physical Properties of Starting Materials

Property	*Virgin Bitumen*	*Maltenes*	*Asphaltenes*	*Weighted Average of Maltenes and Asphaltenes*	*Asphaltene-Enhanced Bitumen*[a]
Carbon (wt %)	86.3	86.7	84.4	86.2	86.1
Hydrogen	11.1	11.7	9.2	11.2	10.8
Nitrogen	0.80	0.64	1.6	0.84	0.90
Sulfur	0.35	0.32	0.56	0.37	0.38
Oxygen	1.4	0.64	4.2	1.4	1.8
H/C atomic	1.53	1.61	1.30	1.55	1.50
Specific gravity 20/20	1.003	0.980	NA[b]	NM[c]	1.018
API gravity	9.5	12.9	NA[b]	NM[c]	7.5
Average molecular weight (VPO benzene)	778	636	8000	2150	NA[b]
Average molecular weight (VPO pyridine)	854	615	1160	727	NA[b]
Con carbon	14.1	8.2	NA[b]	NM	18.0
Asphaltene content	20.6	0	100	20.6	31.1

[a]Elemental analysis and asphaltene content from results obtained on components.
[b]NA—not available.
[c]NM—not meaningful.

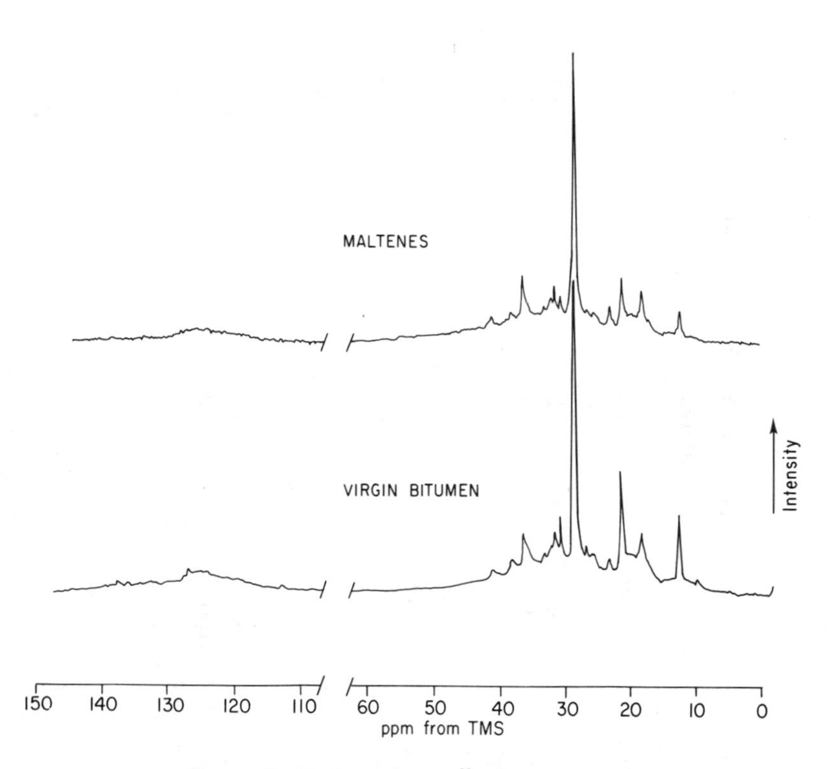

Figure 1. High resolution ^{13}C NMR spectrum

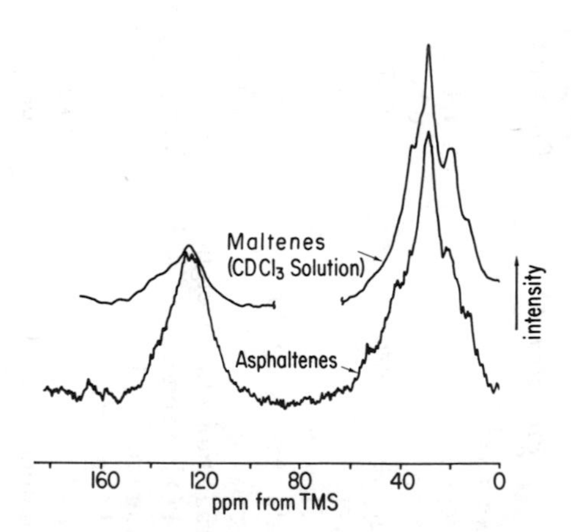

Figure 2. ^{13}C NMR spectra of solid asphaltenes and maltenes

Table II. Semiquantitative Results of ^{13}C NMR

	Total Aromatic Carbon	*Total Saturate Carbon*
Maltenes	17	83
Virgin bitumen	19	81
Asphaltene	33	67

acquired to fully ensure that nonprotonated carbons are not saturated by a too rapid repetition period. Any error due to this effect would be to reduce the apparent concentration of aromatic carbon; however, this effect will be generally less than 25% of the total aromatic carbon under the conditions run. The results indicate that for the bitumen and maltenes only about one carbon in six is bound in an aromatic ring, while for the asphaltenes the proportion is about one in three. Even with the higher aromaticity, the asphaltenes fraction consists predominantly of saturated carbon and any theoretical consideration of structure or processability must take this factor into account. It is apparent from the elemental analysis and ^{13}C NMR data that while the asphaltenes are of a higher molecular weight, aromaticity, and heteroatom content than the maltenes, the gross hydrocarbon structure is not significantly different between these two fractions.

Compound type analysis was not conducted in this work, but it is instructive to look at some literature results. With respect to heteroatoms, tar sand bitumen and petroleum asphaltenes have been variously reported as containing predominantly polar heteroatoms, principally oxygen types (*18*) or nonpolar heteroatoms, principally nitrogen types (*14*). The difference in these reported results apparently relates to the method of analysis in which the former is a direct determination, the latter an indirect determination.

Central to an understanding of the composition of complex hydrocarbon mixtures is that all spectroscopic measurements, elemental analyses, and physical properties represent average values and that a distribution about the average exists. As in many cases, the distribution about an average value is often wide compared with the differences in averages between a maltenes fraction and asphaltenes fraction. Thus, there is frequently broad overlap, and for any given value in the vicinity of the average there will be species in both the asphaltenes and maltenes possessing essentially identical characteristics.

Simulated Distillation Results. Important insight concerning the molecular-size distribution of asphaltenes vs. maltenes is gained by the simulated distillation data. The boiling point distribution curves are shown in Figure 3. These curves were drawn as follows. Quantitative simulated distillation data was obtained on the virgin bitumen and the maltenes. Direct information is obtained up to a nominal boiling point of 535°C shown by the vertical dashed line. The area under the curve for the nonvolatile portion is

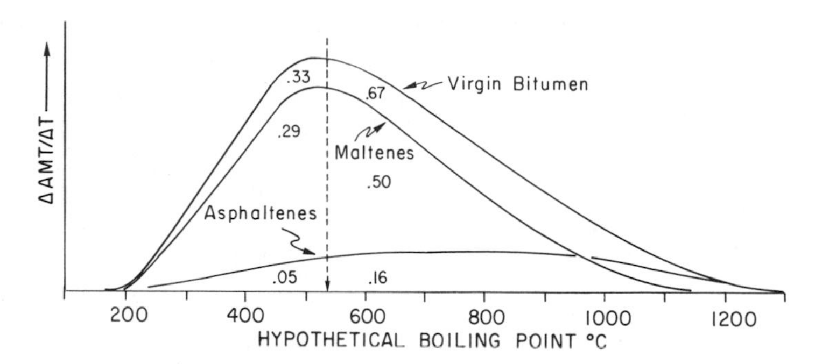

Figure 3. Simulated distillation boiling point curves for virgin bitumen, maltenes, and asphaltenes

determined directly from the simulated distillation data. The end point for the nonvolatile portion of the curve was approximated by knowing the area under the curve and by using molecular weight data to estimate the mean boiling point (this step admittedly requires some judgment). The curves were drawn in a smooth fashion rather than some bimodal configuration because, from the standpoint of geochemical maturation, a distribution with a single maximum is most reasonable. The asphaltenes curve is the difference between the other two curves. Decimal values are the weight fraction of virgin bitumen represented under the curve.

Attempts to obtain a distillation curve directly on the asphaltenes fraction resulted in essentially no FID response. This result is attributed to the high free energy of association of these molecules. Once these species are allowed to associate, it is difficult to disassociate them. The results in Figure 3 suggest that volatile molecules are contained in the asphaltene fraction and that the overlap in molecular weight and volatility between asphaltenes and maltenes is substantial.

Hydropyrolysis Results. The maltenes, virgin bitumen, and bitumen-enhanced to 1.5 times the normal asphaltene content were processed by hydropyrolysis at 500°C, 2000 psig hydrogen pressure, and 30-s gas residence time. The process conditions were selected on the basis of earlier results (*10–12*) for tar sand bitumens, which indicated that gas make and the propensity to form coke increased with increasing temperature, and that a minimum hydrogen pressure of 1200–1500 psig was required to inhibit coking. A moderate temperature of 500°C and, correspondingly, a long residence time was selected in an attempt to accentuate differences in results for the three feeds. Hydropyrolysis is particularly useful for studying asphaltenes since no coke is formed and the largest, most aromatic molecules are retained in liquid products where they can be observed and characterized.

Gravimetric results of the process yields are given in Table III. The percentage of hydrocarbon gases was calculated to close the material balance on carbon and the calculated nonhydrocarbon products were derived by closing the material balance on the heteroatoms. The inclusion of the missing heteroatoms as NH_3, H_2S, and H_2O is done only to provide a basis for calculating the maximum theoretical hydrogen consumption and is not meant to indicate that these products were actually measured.

The results in Table III show that the virgin bitumen that contains the asphaltenes produced relatively more gas and nonhydrocarbon products than did the maltenes. This trend with respect to gases and liquids appears to be confirmed by the results of the run with the asphaltene-enhanced bitumen; however, appreciable quantities of coke were formed at the reaction conditions used and good material balances on this run were not achieved. Without essentially complete reduction of coke formation by hydropyrolysis, the significance of results for the asphaltene-enhanced bitumen are suspect. Removal of carbon in the form of coke will have an unknown effect on results that may not be attributable to asphaltenes. These results are included principally as negative results to show the dramatic effects that can result if asphaltenes are not fully dispersed and coke formation is not inhibited during hydropyrolysis.

In an attempt to quantify the effect of asphaltenes, a calculation that synthesized results of a hypothetical feed of 100% asphaltenes was made based on the maltene and virgin bitumen results. This calculation is useful to break out the relative effect of asphaltenes and maltenes to the total results. The calculation is based on the general relationship: 0.794 (maltene results) + 0.206 (asphaltene results) = virgin bitumen results. Results of this calculation show that asphaltenes have a marked effect on increasing the gas production, and correspondingly, require appreciable quantities of hydrogen for conver-

Table III. Hydropyrolysis Yields

	Maltenes[a]	*Virgin Bitumen*[a]	*Asphaltene-Enhanced Bitumen*[a]	*Asphaltene (Hypothetical)*[a,b]
Hydrocarbon gases	20.1	27.3	NA[d]	54.6
Recovered hydrocarbon liquids	80.3	73.9	42	48.8
Calculated nonhydrocarbon products				
NH_3	0.17	0.38	NA[d]	1.16
H_2S	0.13	0.13	NA[d]	0.13
H_2O	0.46	1.15	NA[d]	3.78
Coke[c]	<1	<1	5–11	NA[c]
	101.2	102.9	NA[d]	10Y5
Calculated hydrogen consumption (wt %)	1.2	2.9	NA[d]	8.5

[a]Weight percent based on feed.
[b]Values for asphaltenes were calculated based on the weighted difference between results for the maltenes and the virgin bitumen.
[c]Small amounts of coke that may be formed from the maltene and virgin bitumen feeds are counted as gases in the calculations.
[d]NA—not available.

sion. The prospect that coke laydown, which may contain mineral matter, on the walls of the reactor catalyzes gas-forming reactions cannot be entirely discounted. It was observed that the small amount of coke formed (less than 0.5% of feed) was not unusually different between the maltene feed and the virgin bitumen feed.

Characterization of Hydropyrolysis Products. The liquid products from hydropyrolysis were characterized by elemental analysis, physical properties, simulated distillation, and ^{13}C NMR. Gaseous products were analyzed by gas chromatography. Characteristics of the liquid products are given in Table IV. Results show products from the maltenes and virgin bitumen are of similar H/C ratios but that the products from the virgin bitumen are of a lower gravity, higher viscosity, and higher molecular weight. The liquid products from the asphaltene-enhanced bitumen do not follow the same trend and, in fact, closely resemble the results obtained on coker distillates from virgin bitumen (*12*). Both the liquid products from the asphaltene-enhanced bitumen and typical coker distillates are amber in color, whereas typical hydropyrolysis liquids are black. The asphaltene-enhanced bitumen has apparently been rendered sufficiently nonvolatile at process conditions that some liquids reside in the reactor for long periods of time (certainly more than the 30-s calculated gas-residence time) and thermally coke. These results can be partially explained by the prospect that the procedure for adding asphaltenes to virgin bitumen failed to redisperse the asphaltenes. The fact that products from the virgin bitumen that contained "in-situ asphaltenes" show little characteristic of a coker distillate indicates that the influence of asphaltenes on the process results of virgin bitumen is mainly attributable to the chemical structure of the asphaltenes and not appreciably to the physical influence of asphaltenes on properties such as volatility and viscosity, and the secondary effects of these properties on chemical reactions. The results of the asphaltene-enhanced bitumen should be interpreted as negative results illustrating the anomalous properties of asphaltenes once they are produced in an isolated fraction.

The "asphaltene contents" of the products were measured and reveal that some 11% asphaltenes are contained in the products from maltene processing (Table IV). The presence of pentane insolubles in this product is remarkable as they are either produced by hydropyrolysis or they are fragments of previously soluble molecules in the maltenes. Later discussion will show that hydropyrolysis successfully inhibits the generation of substantial quantities of aromatic carbon; thus we might reasonably assume the latter explanation dominates.

Perhaps one of the more significant results of this work can be seen in comparison of asphaltenes produced from maltenes and those from virgin bitumen. Taking into account the condensable product yields, it can be shown that hypothetical yield of product asphaltenes from precursor asphaltenes is only about double that of asphaltenes yield from maltenes [e.g., by the previous method of calculation: $0.794\ (0.11 \times 0.803) + 0.206\ (X) = 0.15$

Table IV. Characteristics of Hydropyrolysis Liquid Products

	Feedstock		
Product Properties	*Maltene*	*Virgin Bitumen*	*Asphaltene-Enhanced Bitumen*
Carbon	87.4	86.9	86.5
Hydrogen	11.3	11.4	12.1
Nitrogen	0.63	0.69	0.55
Sulfur	0.25	0.32	0.28
Oxygen	0.29	0.53	0.42
H/C (atomic ratio)	1.54	1.56	1.66
Average molecular weight (VPO benzene)	305	352	272
Specific gravity 20/20	0.942	0.967	0.915
API gravity	18.7	14.8	23.2
Viscosity 20°C (cp)	254 ± 4	2605 ± 50	28 ± .5
Asphaltenes (wt %)	11	15	trace

(0.739); $X = 0.20$). Clearly, a direct chemical relationship between precursor asphaltenes and product asphaltenes does not exist and discussions of conversion of asphaltenes are meaningless. The relative proportion of product asphaltenes from each source is in some agreement, however, with the proportion of aromatic carbon present in the feeds as shown in Table II.

In this regard, it is instructive that the product asphaltenes are highly hydrogen deficient (H/C ~ 0.9), and that there is a marked concentration of nitrogen (3–5%). An H/C ratio of 0.9 is characteristic of an aromatic hydrocarbon possessing about one methyl substituent for every condensed ring. Species with these structural features would be expected to have a high degree of symmetry and little steric interference to π–π associations, thus promoting a favorable free energy of association. The average molecular weight of the product asphaltenes from maltenes is 1100 as measured in pyridine. This high molecular weight is almost surely attributable to intermolecular association since hydropyrolysis is thought to successfully inhibit production of species of higher molecular weight than the reactants (*9, 10, 13*).

Gas Analysis. The hydrocarbon gas analysis is given in Table V. From Table III the gases represented 20, 27, and about 50% of the feed for the maltene, virgin bitumen, and asphaltene-enhanced bitumen feed, respectively. Thus, on a feed basis as asphaltene content increases there is a notable increase in the yield of C_1 and C_2 and a relative decline in C_3, C_4, and C_5. There seems to be no reliable trend in olefin-to-paraffin ratio with increasing asphaltene content; the data are somewhat scattered in this regard. A hypothetical contribution of the asphaltenes to the gas yield is calculated from the maltene and virgin bitumen run and shows the magnitude of the effect of asphaltenes. The negative values for C_4s and C_5s indicate that the asphaltenes not only fail to generate appreciable quantities of these gases but appear to inhibit their formation from the maltenes. Small amounts of CO_2 and CO were detected in all of the gas samples but are not included in the analysis. The accurate measurement of H_2S, NH_3, and other polar gases is also difficult because of the use of a water trap and because of the possible effects of neutralization reactions between acidic gases and basic gases. Failure to account for these species may be significant when attempting to quantify the effect of asphaltenes.

^{13}C NMR of Products. A ^{13}C NMR spectra of the liquid products obtained by hydropyrolysis of the virgin bitumen show that the saturates portion does not differ appreciably from the virgin bitumen. There may be relatively more straight-chain alkyl groups and alkanes in the hydropyrolysis products. The gross similarity of products with feedstocks has been observed previously with bitumens and coker distillates (*19*) and indicates that major structural changes are not effected by the process. Recognizing that ^{13}C NMR spectra are somewhat insensitive to minor changes, if major structural changes were taking place we might observe a more pronounced disappearance of one type of saturated carbon relative to another, more stable type.

Spectra obtained on the liquid hydropyrolysis products under identical conditions to those for the virgin bitumen revealed an aromatic carbon content of 30%. It is instructive to attempt to determine the fate of the original aromatic carbon present in the bitumen. Assuming that all of the aromatic carbon appears in the liquids, and assuming that ^{13}C NMR data obtained under identical conditions are subject to about the same magnitude of error for similar samples, the total aromatic carbon in the hydropyrolysis products can be calculated to be about 22% (30% × 0.739). This compares with the original 19% and suggests that hydropyrolysis inhibits the formation of large quantities of aromatic carbon. This is particularly notable in light of the high concentration of naphthenic rings, which may be expected to dehydrogenate through hydrogen-transfer reactions under the conditions studied. For example, in coking, assuming all of the carbon in coke to be aromatic carbon, the total

Table V. Hydrocarbon Gas Analysis

	Feedstock			
	Maltenes	*Virgin Bitumen*	*Asphaltene-Enhanced Bitumen*	*Asphaltenes (Hypothetical)*
Methane	23.5	29.7	33.1	53.6
Ethylene	2.9	2.9	2.7	2.9
Ethane	10.6	12.8	15.0	21.3
Propylene	10.9	13.8	12.5	25.0
Propane	17.1	14.6	14.3	5.0
Isobutane	3.0	2.3	2.1	−0.4
n-Butenes	7.6	5.2	3.6	−4.1
n-Butane	10.0	7.5	6.9	−2.1
Isobutylene	0.9	0.7	0.5	−0.1
C_5's	13.5	10.5	9.3	−1.1
	100	100	100	100

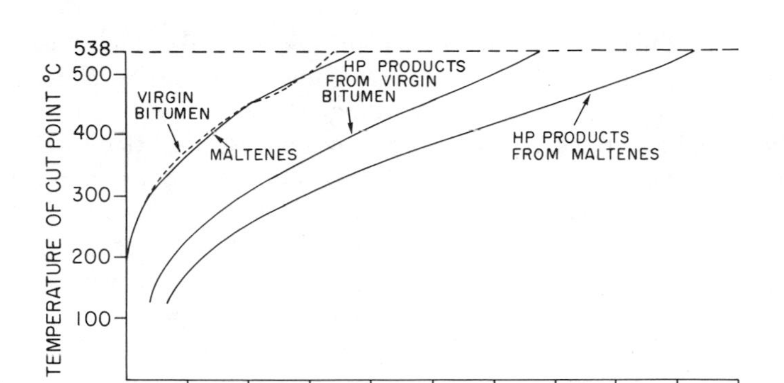

Figure 4. Simulated distillation boiling point curves for feedstocks and products

aromatic carbon in products is 30% for a Sunnyside bitumen exhibiting substantial dehydrogenation.

Simulated Distillation of Products. The simulated distillation curves of the maltene and virgin bitumen feedstocks, along with their respective hydropyrolysis products, are given in Figure 4. In contrast to the feedstock materials, the products reveal appreciable differences. In this regard, the possibility exists that actual reactor residence times (liquid phase) differ somewhat and influence the results. Neglecting this possibility, however, it appears that the asphaltenes are not as easily converted to distillable products, and the conversion products from the asphaltenes generally make up the nondistillable liquids and the gases. This observation is consistent with the yield results discussed previously.

Discussion

The chemical structure and conversion characteristics of petroleum asphaltenes have been a subject of longstanding interest. A traditional viewpoint of asphaltenes has been that the molecules representing this fraction are of unusually high molecular weight, high in aromaticity, high in heteroatom content, and some way structurally unique from nonasphaltene species. Some publications have suggested that the high molecular weight derives from the presence of macromolecules possessing large clusters of condensed ring systems (*see,* for example, References *20* and *21*). Others have suggested that the unit structures are of a more moderate molecular weight but are polymerized through carbon–carbon bonds (*22, 23*) or sulfur linkages (*24, 25*). The effect of solvent polarity on measured molecular weights (*26*) and the possible role of peptization of asphaltenes by resins have been reported for bitumen (*27*).

For the most part, the differences between various models and viewpoints alluded to above can be reduced to differences in opinion as to the true molecular weight of individual molecules. Because the molecular-weight distribution and average is central to an understanding of the chemical structure, this point deserves some discussion.

One of the authors has previously postulated (*14*) that the average molecular weight for petroleum asphaltenes is in the upper hundreds rather than thousands. This contention is supported by several arguments. First, from the standpoint of geochemical maturation, considering the prospects for adsorption during migration, it is not reasonable that macromolecules are both formed and remain unabsorbed to be produced with the oil. Second, when total oils are examined in the absence of solvent, extremely high molecular-weight species do not seem to be present. For example, Weeks (*28*) observes by mass spectrometry an average molecular weight of only 300–400 for the heaviest fraction obtained by ultracentrifugation. Third, no study has unambiguously established an average molecular weight for petroleum-derived asphaltenes, the lowest achieved by measurement being in the low thousands. These measurements and those by GPC are clouded by the prospect of an unknown degree of intermolecular association. Fourth, generally all of the properties attributable to asphaltenes can be explained with average molecular weights in the hundreds rather than thousands, for example, it is not necessary to invoke molecular weights above 1000 to explain the properties observed. McKay et al. (*18*) have also contended that petroleum asphaltenes are monomeric in nature and possess average molecular weights in the range of 500–800.

Common to many arguments of asphaltene structure is the tendency to take the gross or average properties and project from these values some average structure with little recognition of the width of the distribution about the average. Such representations always are misleading and often are erroneous. Unless the distribution of molecular species, with respect to molecular weight, volatility, aromaticity, heteroatom content or other property, is taken into account, an accurate representation of asphaltenes cannot be made. Models based on average properties, even if based on accurate data, can at best represent only a small fraction of the species present.

Recognition of the distribution(s) inherent in petroleum, bitumens, or asphaltenes leads to some general concepts regarding the chemistry of asphaltenes. First, and speaking strictly from an operational standpoint, asphaltenes differ from nonasphaltenes only in their solubility properties under the conditions of solvent deasphaltening. Under equilibrium conditions (which may be approached during 24-hr digestion) solution thermodynamics state that the chemical potential (partial molar-free energy, $\Delta\overline{G}_i$) of a given species in the solid (asphaltene) phase must be equal to the chemical potential of that species in the solution (maltene) phase. The partial molar-free energy can be further related to the partial molar enthalpy and entropy (which can be related to structure as discussed below) by the expression, $\Delta\overline{G}_i = \Delta\overline{H}_i - T\Delta\overline{S}_i$. Because $\{\Delta\overline{G}_i\}^{\mathrm{I}} = \{\Delta\overline{G}_i\}^{\mathrm{II}}$, where I and II denote the two phases, a proportion of species (i) will appear as an asphaltene and a proportion will appear as a maltene, the magnitude of the proportions depending upon the relative magnitude of the enthalpies and entropies in the two phases. Thus, thermodynamic considerations predict that for those species of borderline solubility, species will be found in both phases. For complex hydrocarbon systems, such as bitumens for which the geochemical maturation process has generated a broad continuum of species with respect to properties such as molecular weight and aromaticity, it can reasonably be assumed that a substantial portion of a bitumen possesses the appropriate thermodynamic properties to be distributed to an appreciable extent in both the asphaltene phase and the solution phase. Scheppele (*29*) presents strong evidence by mass spectrometry that identical species are indeed found in both the asphaltenes and maltenes, although proportions vary from species to species.

Second, the compound types and molecular symmetries that determine the relative magnitudes of the partial molar enthalpies (electronic factors) and entropies (geometric factors), respectively, vary broadly with respect to molecular weight. Thus, low-molecular-weight polar or highly aromatic species may appear as asphaltenes while larger naphthenic-aromatics may appear as maltenes. This concept has been developed by Long (*30*) and by McKay et al. (*18*). Considering the molecular structure of bitumen (*2, 31*) and petroleum heavy ends (*32–34*), which show that aromatics and polar heteroatoms are broadly distributed with respect to molecular weight, substantial overlap with respect to molecular-weight distribution is expected to exist between the maltenes and the asphaltenes. This factor is confirmed by results depicted in Figure 3.

The effect of these two distributions, namely the partitioning of species between the two phases, and the overlap of the two phases with respect to molecular weight and functionality implies that the separation between maltenes and asphaltenes is nondefinitive in terms of chemical functionalities and structures. While properties measurements commonly detect differences in the average properties, these differences are small when compared with the width of the distribution about that average.

This concept of asphaltenes is useful in the interpretation of the present data, and conversely, the data support the concept. First, ^{13}C NMR data show that the saturated hydrocarbon structure, which constitutes the majority of the carbon in the fractions, is virtually identical between the asphaltenes and the maltenes, within the limited sensitivity of ^{13}C NMR. This factor is consistent with the argument that there is a partitioning between fractions and that the appearance of a particular species predominantly in the asphaltene fraction results because of a relatively higher aromaticity or the presence of polar heteroatoms for a specified molecular weight. It is important to recognize, from a processing standpoint, that only a minor weight percent of the fraction (or molecule) may be responsible for its classification as an asphaltene.

Second, the simulated distillation data, which show the maltenes to be 35–40% volatile while the asphaltenes are about 20–25% volatile, indicate that the asphaltenes and the maltenes are widely distributed with respect to volatility. This observation is particularly important because the maltenes must contain very high-molecular-weight material while the asphaltenes include significant quantities of low-molecular-weight material. This is semiquantitatively shown in Figure 3.

Third, the measured molecular weight of the virgin bitumen does not correspond to the weighted average molecular weight of the parts. These data illustrate that the disproportionate errors encountered are probably attributable to intermolecular association. Considering that even saturated hydrocarbons of an average molecular weight of 325 associate in benzene and give erroneously high VPO molecular weights by 12–15% (*14*), it is reasonable to assume that even the maltene and virgin bitumen measured molecular weights are erroneously high by at least 20–30% of the true values. Molecular weights of asphaltenes calculated by extrapolation of the maltenes and virgin bitumen data place the value for asphaltenes in the upper hundreds.

Fourth, the result of hydropyrolysis reactions are consistent with the concept in several important respects. That products from the maltene hydropyrolysis contain appreciable quantities of asphaltene despite the low net production of aromatic carbon indicates that high-molecular-weight aromatics found in the maltenes crack by a similar mechanism, leaving aromatic nuclei that are rendered insoluble in *n*-pentane by loss of alkyl substituents. As was pointed out above, a direct relationship between virgin asphaltenes and product asphaltenes apparently does not exist. Just as product asphaltenes are derived from maltene precursors, asphaltene precursors contribute significantly (i.e., 80%) to gaseous and nonasphaltene liquid products. These results strongly suggest that empirical relationships between asphaltene content and difficulty with processability, while useful to the process engineer, are chemically fortuitous.

The approach of studying asphaltenes indirectly has resulted in observations vis-à-vis, volatility, processability, and molecular weight that are not normally observed when examining isolated fractions of asphaltenes. We have speculated that this is because of the high free energy of association incurred when asphaltenes are precipitated. This phenomenon explains apparent discrepancies in the literature that show asphaltenes to be primarily acidic (*18*) when studied directly, and primarily basic or neutral functionality when studied indirectly (*12, 14*). While the conclusions of this study may not generally apply to petroleum asphaltenes of all origins, the arguments relating to partitioning and overlap in distribution are felt to apply to asphaltenes of any origin.

Conclusions

The structure and chemistry of tar sand bitumen asphaltenes were studied indirectly by inference from processing and characterization of virgin bitumen and deasphaltened bitumen (maltenes). Results differ significantly from those derived from characterization of asphaltenes as isolated from bitumen. Hydropyrolysis is particularly well-suited as a process reaction for studying the chemistry of asphaltenes because formation of coke and high-molecular-weight species are inhibited and converted asphaltenes contribute entirely to analyzable products.

Molecules comprising the asphaltene fraction from bitumen are suggested to have average molecular weights in the mid- to high-hundreds and are broadly distributed with respect to molecular weight, aromaticity, and heteroatom content. Molecules comprising the maltene fraction, while of a lower average molecular weight, aromaticity, and heteroatom content, are also broadly distributed. There is substantial overlap between asphaltenes and maltenes with respect to these chemical functionalities and molecular weight. This conclusion is supported by the results of the hydropyrolysis conversion and simulated distillation data. In addition, theoretical arguments, supported by literature evidence, suggest that a partitioning of any given species occurs between the maltenes and asphaltenes. The results of this study indicate that asphaltenes are chemically nondefinitive in terms of compound types, functionality, or other structural features. On this basis, correlations between asphaltenes content and processability are probably fortuitous. It is suggested that if conversion chemistry of high-molecular-weight fractions is to be elucidated, the concept of asphaltenes should be abandoned, and more definitive approaches using compound type analysis by chromatographic and spectroscopic analysis of total feedstocks should be employed instead.

Acknowledgements

The assistance of Dr. K. W. Zilm in obtaining and assisting in interpretation of ^{13}C NMR spectra is appreciated. The financial support of the State of Utah and U.S. Department of Energy is gratefully acknowledged.

Literature Cited

1. Bunger, J. W. In "Shale Oil, Tar Sands, and Related Fuel Sources," *Adv. Chem. Ser.* **1976,** *151*, 121.
2. Bunger, J. W.; Thomas, K. P.; Dorrence, S. M. *Fuel* **1979,** *58*(3), 183.
3. Zilm, K. W.; Pugmire, R. J.; Grant, D. M., unpublished data.
4. Cross, V. R.; Hester, R. H.; Waugh, J. S. *Rev. Sci. Instrum.* **1976,** *47*, 1186.
5. Zilm, K. W.; Alderman, D. W.; Grant, D. M. *J. Magn. Reson.* **1978,** *30*, 563.
6. Pines, A.; Gibby, M., Waugh, J. S. *J. Chem. Phys.* **1973,** *59*, 569.
7. Hartmann, S. R.; Hahn, E. L. *Phys. Rev.* **1962,** *128*, 2042.
8. Zilm, K. W.; Pugmire, R. J.; Grant, D. M.; Wood, R. E.; Wiser, W. H. *Fuel* **1979,** *58*(1), 11.
9. Shabtai, J.; Ramakrishnan, R.; Oblad, A. G. In "Thermal Hydrocarbon Chemistry," *Adv. Chem. Ser.* **1979,** *183*, 297–328.
10. Ramakrishnan, R., Ph.D. Dissertation, Univ. of Utah, 1978.
11. Bunger, J. W.; Cogswell, D. E.; Oblad, A. G. *Am. Chem. Soc., Div. Fuel Chem., Prepr.* (Miami Beach, Sept., 1978) *23*(4), 98.
12. Bunger, J. W. Ph.D. Dissertation, Univ. of Utah, 1979.
13. Bunger, J. W.; Cogswell, D. E.; Wood, R. E.; Oblad, A. G.; In "Oil Shale, Tar Sands and Related Materials," *A. C. S. Symp. Ser.* **1981,** *163*, 371–382.

14. Bunger, J. W. *Am. Chem. Soc., Div. Pet. Chem., Prepr.* (New Orleans, March, 1977) *22*(2), 716.
15. Speight, J. G. "The Structure of Petroleum Asphaltenes—Current Concepts Information Series—81," Alberta Research Council, 1978.
16. Moschopedis, S. E.; Speight, J. G. *Fuel* **1975,** *54*, 210.
17. Moschopedis, S. E.; Fryer, J. F.; Speight, J. G. *Fuel* **1976,** *53*(3) 184.
18. McKay, J. F.; Amend, P. J.; Cogswell, T. E.; Harnsberger, P. M.; Erickson, R. B.; Latham, D. R. In "Analytical Chemistry of Liquid Fuel Sources," *Adv. Chem. Ser.* **1978,** *170*, 128.
19. Bunger, J. W., unpublished data.
20. Yen, T. F. *Am. Chem. Soc., Div. Pet. Chem., Prepr.* (New York, Aug., 1972) *17*(4), F102.
21. Witherspoon, P. A. In "Fundamental Aspects of Petroleum Geochemistry," Nagey, B.; Colombo, A. Eds. Elsevier: New York, 1967; Chap. 6.
22. Mitchell, D. L.; Speight, J. G. *Fuel* **1973,** *52*(2), 149.
23. Yen, T. F. Chapter 4 in this book.
24. Ignasiak, T.; Bimer, J.; Samman, N.; Montgomery, D. S.; Strausz, O. P. Chapter 11 in this book.
25. Ignasiak, T.; Kemp-Jones, A. V.; Strausz, O. P. *J. Org. Chem.* **1977,** *42*(2), 312.
26. Moschopedis, S. E.; Fryer, J. F.; Speight, J. G. *Fuel* **1976,** *55*(3), 227.
27. Koots, J. A.; Speight, J. G. *Fuel* **1975,** *54*(3), 179.
28. Weeks, Jr., R. W.; McBride, W. L., *Am. Chem. Soc., Div. Petr. Chem. Prepr.* (D.C., Sept., 1979) *24*(4).
29. Scheppele, S. E.; Benson, P. A.; Greenwood, G. J.; Aczel, T.; Grindstaff, Q. G.; Beier, B. F. Chapter 5 in this book.
30. Long, R. B. Chapter 2 in this book.
31. Selucky, M. L.; Chu, R.; Ruo, T.; Strausz, O. P. *Fuel* **1977,** *56*(4), 369.
32. McKay, J. F.; Cogswell, T. E.; Weber, J. H.; Latham, D. R. *Fuel* **1976,** *54*(1), 50.
33. Dooley, J. E.; Hirsch, D. E.; Thompson, C. J.; Ward, C. C. *Hydrocarbon Process.* **1974,** *53*(11), 187.
34. Jewell, D. M.; Weber, J. H.; Bunger, J. W.; Plancher, H.; Latham, D. R. *Anal. Chem.* **1972,** *44*(9), 1391–1395.

RECEIVED October 6, 1980.

Composition of Asphaltenes from Coal Liquids

THOMAS ACZEL, R. B. WILLIAMS, N. F. CHAMBERLAIN, and H. E. LUMPKIN

Exxon Research and Engineering Company, Baytown, TX 77520

Asphaltenes from heavy coal distillates were characterized with a combination of high resolution mass spectrometry, nuclear magnetic resonance, infrared, gel permeation chromatography, and elemental analyses. More than 1000 aromatic and heterocyclic carbon number components were identified and determined in one sample alone. Structural assumptions were cross-checked by comparing average values on carbon atom types calculated from the MS composition with averages obtained directly from NMR. Coal asphaltenes consist predominantly of units of three to ten ring heteroaromatic components, with one to two or three of these units per molecule. Most abundant heteroaromatic ring systems are those with two oxygen atoms per molecule, followed by monooxygen, trioxygen, nitrogen, nitrogen oxygen, sulfur oxygen, and hydrocarbon type units. Sidechains and naphthenic rings are also present, with one or two naphthenic rings and one or two short sidechains associated with each heteroaromatic or aromatic nucleus.

Asphaltenes are generally defined as those components in petroleum and coal liquids that under certain conditions are soluble in benzene but insoluble in aliphatic solvents, such as *n*-pentane, *n*-heptane, or cyclohexane. This definition obviously includes a broad variety of components, as insolubility in the above solvents can be caused by high molecular weight, high polarity, hydrogen bonding, acid–base complexing, or combinations of these parameters.

Asphaltenes are thought to be a major intermediate in the thermal processes leading to coal liquefaction. Increased interest in coal liquefaction techniques has, therefore, resulted in renewed attempts to unravel the structural characteristics of these materials. Fortunately, this effort is facilitated by

0065-2393/81/0195-0237$05.00/0

the very nature of coal-derived asphaltenes, which are generally more polar but of lower molecular weights than petroleum asphaltenes, and, thus, are more amenable to detailed compositional analyses.

We have recently determined the structural parameters and composition of some asphaltene samples obtained from the Synthoil and Exxon Donor Solvent (EDS) liquefaction processes. The particular EDS sample used was sufficiently volatile for analysis by ultrahigh resolution mass spectrometry, so we could obtain very detailed data on its composition in terms of the distribution of individual carbon-number homologs. Information from this approach, integrated with data from NMR, IR, molecular weight determinations, elemental analyses, and separations furnished us with a novel and detailed insight into the nature of these asphaltenes. The excellent agreement observed between composites calculated from the detailed MS data, where available, and the averages determined by NMR, IR, and elemental analyses reinforces the credibility of the approaches used and allows extrapolations to heavier samples that are not amenable to detailed MS characterization.

This chapter describes briefly the analytical methodology used in this work, summarizes the more than 1200 individual asphaltene components determined, and discusses the overall structural implications of the integrated analytical findings.

Discussion

Analytical Methodology. Analytical techniques used in this work included separations, high and low resolution MS, NMR, IR, UV, molecular weight determinations, and elemental analyses. These are discussed in detail in our work on the chemical characterization of Synthoil feeds and products (*1*, *2*). Composition of the EDS asphaltenes was determined with the aid of a Kratos-AEI model MS50 ultrahigh resolution mass spectrometer that became available only after we had already completed the work on the Synthoil samples. The roles of the various analytical approaches selected are listed in Table I. The most important techniques are also summarized in the following paragraphs.

SEPARATIONS. The asphaltene fractions were obtained by solvent extraction with benzene and subsequent precipitation with cyclohexane. The cyclohexane-soluble fractions were separated into saturate, aromatic, and polar aromatic fractions by the clay–gel technique, ASTM D-2007 (modified). This separation is also applicable to asphaltenes.

HIGH RESOLUTION MASS SPECTROMETRY. High resolution mass spectrometry was used to identify the maximum possible number of components, or rather, carbon-number homologs, in the samples. This can be achieved only on compounds that are volatile at the conditions used (300° C and 1×10^{-2} torr) and yield stable molecular ions. These conditions are satisfied by esentially all aromatic and heteroaromatic components boiling below approximately

Table I. Roles of Major Analytical Techniques in Characterization of Asphaltene Fractions

Technique	*Examples of Structural Information*
Solvent extraction[a]	Concentration of asphaltenes in sample
Chromatographic separation[a]	Saturates/aromatics/polar aromatics
Low resolution mass spectrometry	Residual solvents, volatiles, initial survey of sample nature, MW distribution, determination of saturate types
Ultrahigh resolution mass spectrometry	Determination of individual aromatic and heteroaromatic carbon-number homologs. Calculation of composite structural features from MS composition
Nuclear magnetic resonance	Determination and calculation of average structural parameters
Infrared spectrometry	Functional groups (OH, COOH, $CONH_2$)
Ultraviolet/visible spectrometry	Aromatic ring structures
Miscellaneous conventional techniques and neutron activation analysis	C, H, O, N, basic N, S
Gel permeation chromatography and vapor pressure osmometry	Average molecular weights, molecular weight distribution

[a] Used also to prepare fractions for subsequent additional separations and/or instrumental analysis.

600° C and likely to be present in coal liquids. In the case of the EDS asphaltenes, these amounted to 77% of the sample; in case of the Synthoil asphaltenes, to 10%–30%.

The mass spectrometer was operated in the low voltage mode at a resolving power of about 50,000. This mode of operation restricts the mass spectra to aromatic and heteroaromatic molecular ions and allows the separate determination of essentially all volatile hydrocarbons, aromatic oxygen, and aromatic nitrogen compounds that are present in coal liquefaction products. The high accuracy of the computerized mass measurements (better than 0.0005 ama units on the average) allows the identification of some of the sulfur-containing components, even if the instrument is operated at a resolving power below that required (approximately 1:90,000 at mass 300) to separate these components from interfering hydrocarbons ($C_3 - SH_4$ doublet; ΔM = 0.0034 ama). The power of ultrahigh resolution mass spectrometry is illustrated by the determination of approximately 1200 components/carbon-number homologs in the spectrum of the EDS asphaltenes.

The MS data are quantitated with the aid of proprietary computer programs. These include programs that calculate composite values from the MS composition for average parameters typically determined by NMR, such as fraction of carbons in aromatic structures and angular H atoms. The scope of these programs is still increasing.

PROTON NUCLEAR MAGNETIC RESONANCE. Proton nuclear magnetic resonance (NMR) and in one case ^{13}C NMR were the major techniques used to determine the average structural features of the asphaltene fractions. Interpretations were based on correlations published in the literature (*3*) and on calculations based on the combination of data from NMR, GPC, and elemental analyses (*4, 5*). These latter are based on equations relating total rings (R), aromatic rings (R_a), naphthenic rings (R_n), total hydrogens (H), total carbons (C), aromatic ring carbons (C_a), peripheral (substitutable) aromatic ring carbons (C_1) and number of alkyl links between aromatic ring groupings (L) (*6*).

$$2(R_a - 1)/C = 2 - H/C - C_a/C$$

$$2(R - 1)/C = (C_a - C_1 + 2(R_n + L))/C$$

A detailed derivation of these equations is given in References *1* and *2*. Their main function is to relate NMR data to atomic carbon types and to convert the relative ratios among these to number values. The calculations yield only average values and include uncertainties, as for example that between alkyl linkages and naphthenic rings (only the sum of these can be calculated). Nevertheless, they can be used to define reasonable structural limits. Particularly interesting is the comparison of these NMR parameters with MS data, because this approach, as will be shown later, allows the deduction of structural parameters that could not be possibly obtained from MS or NMR.

Infrared Spectrometry. Infrared spectrometry was used for the determination of functional groups and for survey purposes. Samples were run both in solution and as KBr pellets. Assignments were based on the work of McKay et al. (*7*), Snyder (*8*), and Petersen et al. (*9*).

A very detailed description of the overall analytical methodology is reported in Reference *1*.

Results

The structural information obtained will be illustrated with data on asphaltenes from a Clearfield, Pennsylvania coal extract, asphaltenes from the Synthoil coal liquefaction process of the Department of Energy, and a lighter distillate asphaltene from the Exxon Donor Liquefaction Process (EDS). No attempt is made to correlate composition with sample origin or treatment as the samples derive from different coals and were obtained at different conditions.

The data yield a considerable and novel insight into the structure of asphaltenes from coal liquids. This is particularly true for the lighter distillate asphaltenes from the EDS process, for which detailed MS data are available. The good agreement observed between data from MS, NMR, and other techniques increases the credibility of the latter even in the cases where the samples were not sufficiently volatile for detailed MS characterization.

Our major findings are summarized below:

1. The asphaltenes studied are extremely complex mixtures of hydrocarbons and heteroatomic compounds.
2. Volatilized asphaltene molecules contain prevalently only one structural unit per molecule. Heavier molecules contain one to three of these units per molecule.
3. Basic structural units are two- to eight-ring condensed aromatic hydrocarbons, associated with one to three functional groups and/or naphthenic rings and short sidechains. Average aromatic ring condensation ranges from eight to ten in the heavier asphaltenes, three to five in the lighter (EDS) asphaltenes. There is no evidence of alkyl links in the latter.
4. Average molecular weight is 500–800.

The above conclusions apply to all asphaltenes studied, although detailed MS characterization was limited to the lighter EDS asphaltene samples. High resolution MS spectra obtained for the 10%–30% volatiles from the other asphaltene samples show the presence of the same type of components as the ones determined in the light asphaltenes.

The high resolution MS analysis of the light asphaltene sample is summarized in Tables II–VI.

The overall distribution of the components determined is shown in Table II. Items reported for each class of components include total amount, the number of homologous series (in Z number), and individual carbon-number homologs determined in that class and their ranges (in carbon number).

The Z number indicates the hydrogen deficiency in a homologous series with general formula C_nH_{2n-Z}. Most abundant components are those with two oxygens atoms per molecule, followed by those with one oxygen atom per molecule, three oxygen atoms per molecule, and hydrocarbons. Some of the miscellaneous components, although present only in very minor amounts, contain as many as five heteroatoms per molecule. In all, we have determined a total of 1269 components, ranging from 30 ppm to 20,000 ppm concentration in this asphaltene fraction and belonging to 234 homologous series. Ring condensation ranges from one to eight (from Z = 6 to Z = 44), and the carbon-number homologs within the series range from C_6 to C_{30}.

The presence of hydrocarbons and of low molecular weight heterocompounds identical to those found in the cyclohexane- or pentane-soluble fractions of coal liquids was rather unexpected. However, neutral hydrocarbons have been identified in other coal asphaltenes (*10*), and more recently,

Table II. Compound Types Determined in Light Asphaltenes From EDS Process

		Homologous Series		*Homologs*	
Class	*Weight Percent*	*Number*	*Range in Z Number*	*Number*	*Range in Carbon Number*
Dioxygen compounds	27.26	20	6–44	130	6–29
Monooxygen compounds	16.81	19	6–42	139	10–30
Hydrocarbons	6.29	18	12–46	101	10–31
Trioxygen compounds	6.26	14	12–38	104	12–28
Nitrogen compounds	5.84	20	5–43	131	9–30
Nitrogen–oxygen compounds	3.57	15	7–39	118	8–29
Thiophenes	2.40	11	10–36	64	8–28
Miscellaneous N, S, O compounds	2.19	69	8–35	231	9–27
Sulfur–oxygen compounds	2.18	12	10–32	41	10–28
Sulfur–dioxygen compounds	1.90	13	10–34	37	10–23
Miscellaneous S, O compounds	1.54	10	10–38	119	13–23
Tetraoxygen compounds	0.96	13	10–34	54	14–24
Nonvolatile residue	22.80	—	—	—	—
Totals	100.00	234	5–46	1269	6–31

we have also separated saturate, aromatic, and polar aromatic fractions from a coal asphaltene sample using the clay gel percolation procedure (ASTM D 2007). We speculate that these neutral molecules coprecipitate with the surrounding polar materials, possibly entrapped within a hydrogen-bond network.

The aromatic ring distribution determined for the major classes of compounds is shown in Table III. Four and five ring types are the most abundant. Hydrocarbons show higher average condensation than the heterocompounds; however, most of the latter contain additional functional groups such as hydroxy groups and furanic rings.

The distribution of homologous series within the major classes is shown in Table IV. It is essentially a Gaussian type of distribution; the most abundant series are those with general formulas ranging from C_nH_{2n-26} to C_nH_{2n-30}, $C_nH_{2n-20}O$ to $C_nH_{2n-30}O$, and $C_nH_{2n-16}O_2$ to $C_nH_{2n-26}O_2$. The sudden increase in concentration for the $C_nH_{2n-16}O_2$ series could be an indication of the presence of hydroxydibenzofurans.

Table V illustrates the carbon-number distribution in selected series. Similar trends were observed in most others. With one exception, the homolog present in the largest concentration within a series has three to four carbon atoms more than the first member of the series. This could indicate short sidechains or, possibly, the appearance of isomeric naphthenoaromatic series

Table III. Aromatic Ring Distribution in Major Classes Found in EDS Asphaltenes

	Weight Percent						
Number of Aromatic Rings/Molecule	*Hydro-carbons*	*Monooxygen Compounds*	*Dioxygen Compounds*	*Trioxygen Compounds*	*Nitrogen Compounds*	*Thiophenes*	*Total*[a]
1	—	0.34	0.40	—	0.22	—	0.96
2	0.93	0.78	4.92	1.28	0.49	0.02	8.42
3	0.71	2.05	9.76	1.70	0.71	0.92	15.85
4	2.09	9.42	9.22	2.59	2.99	0.86	27.19
5	1.85	2.86	2.37	0.59	1.03	0.54	9.24
6	0.67	1.36	0.59	0.10	0.40	0.06	3.18
7–8	0.04	—	—	—	—	—	0.04
Totals	6.29	16.81	27.26	6.26	5.84	2.40	64.86
Average number of aromatic rings/molecule	4.12	4.06	3.36	3.45	3.91	3.88	3.69

[a] Ring distribution in minor unreported classes, amounting to 12.34%, is similar to that shown above; degree of condensation should be higher in 22.80% nonvolatiles.

Table IV. Distribution of Homologous Series in Selected Classes Found in EDS Asphaltenes

Z Number in General Formula, $C_nH_{2n-Z}X$	*Hydro-carbons*	*Monooxygen Compounds* X = *O*	*Dioxygen Compounds* X = O_2
6	—	0.15	0.34
8	—	0.13	0.04
10	—	0.06	0.02
12	0.24	0.13	0.06
14	0.30	0.13	0.68
16	0.38	0.52	4.19
18	0.41	0.81	4.72
20	0.30	1.24	5.02
22	0.68	3.57	4.87
24	0.54	3.93	2.66
26	0.87	1.92	1.71
28	1.29	1.82	1.56
30	0.56	1.03	0.81
32	0.51	0.73	0.43
34	0.07	0.42	0.15
36	0.09	0.22	0.01
38	0.02	0.16[a]	0.003[a]
40	0.01	0.10[a]	0.001[a]
42	0.01	0.04[a]	0.001[a]
44	0.006[a]	—	traces[a]
46	0.004[a]	—	—
Total	6.29	17.11	27.27
Average Z number overall average Z number, 22.49	24.47	24.30	20.82

[a]These types were included in the miscellaneous category in Tables II and III.

starting at higher carbon numbers. Both structural types are probably present. This type of carbon-number distribution is similar to that observed in cyclohexane-soluble coal liquids, and in organics directly volatilized from coal (*1*), and it is an indication of intermediate coal rank and relatively mild temperature conditions in processing.

One of the most salient features of the detailed MS data summarized in Tables II–V is the similarity of the components determined in the light asphaltenes to those generally found in the cyclohexane-soluble fractions. Asphaltenes are simply more polar than the cyclohexane solubles, but less polar than the polar fraction separated from the cyclohexane-soluble fraction. This is attributable to the fact that asphaltenes contain significant amounts of neutral components. The overall amount of polar components is much higher in asphaltenes than in the cyclohexane solubles, and this is the factor that is

Table V. Carbon-Number Distribution in Selected Homologous Series in EDS Asphaltenes

	Weight Percent					
Carbon Number	C_nH_{2n-22}	C_nH_{2n-28}	$C_nH_{2n-22}O$	$C_nH_{2n-28}O$	$C_nH_{2n-22}O_2$	$C_nH_{2n-28}O_2$
16	0.15	—	0.04	—	0.18	—
17	0.08	—	0.32	—	0.66	—
18	0.07	—	0.69	—	0.86	—
19	0.07	—	0.82	—	0.96	—
20	0.10	0.04	0.63	0.27	0.66	0.35
21	0.08	0.18	0.48	0.37	0.55	0.30
22	0.07	0.29	0.28	0.32	0.38	0.18
23	—	0.24	0.13	0.26	0.28	0.20
24	0.02	0.16	0.08	0.21	0.18	0.19
25	—	0.12	0.06	0.18	0.09	0.13
26	0.02	0.05	0.04	0.09	0.04	0.09
27	—	0.04	—	0.06	0.03	0.06
28	0.02	0.13	—	0.02	—	0.04
29	—	0.04	—	0.04	—	0.02
30	—	—	—	—	—	—
Totals	0.68	1.29	3.57	1.82	4.87	1.56

probably responsible for their insolubility at the conditions used. Asphaltenes also seem to have, on the average, more heteroatoms per molecule than the polar components of the cyclohexane-soluble fractions. Components with two oxygen atoms per molecule, for example, are much more abundant than those with one oxygen atom per molecule in asphaltenes, while monooxygen compounds are more abundant than dioxygen compounds in the polar fraction separated from the corresponding cyclohexane solubles (27% and 17%, respectively, in asphaltenes vs. 14% and 26% in the polars). Some of these trends are illustrated in Table VI.

Although MS data can be used to obtain structural information, these data usually need to be extended and corroborated by other analytical techniques such as elemental analyses, molecular weight measurements, NMR, IR, and UV. The remainder of this chapter deals with information gathered from these techniques and with the integration of the data with the above-discussed MS characterization. As volatility is not a limiting factor for these determinations, the data reported include those obtained on the heavier asphaltenes.

Elemental analyses of three selected asphaltenes are reported in Table VII. Comparison of the data from conventional methods, in particular the H/C ratio, confirms the "true" asphaltenic nature of the asphaltene fraction from the EDS distillate. The higher oxygen content of the latter indicates higher polarity, and it might explain its lower molecular weight and higher volatility in the MS inlet (Table VIII). Comparison of conventional analyses on the total sample and its volatile fraction indicates a lower H/C and higher oxygen content for the 23% nonvolatiles.

The molecular weight data given in Table VIII were obtained by gel permeation chromatography using both tetrahydrofuran and *N*-methylpyrrolidone solvents and were confirmed by VPO.

Table VI. Relative Concentrations of Heterocyclics in Aromatic, Polar, and Asphaltene Fractions from EDS Asphaltenes

	Weight Percent		
Compound Class	*Aromatics*	*Asphaltenes*	*Polar Aromatics*
Hydrocarbons	94.58	6.29	2.94
Monooxygen compounds	4.97	16.81	26.39
Dioxygen compounds	0	27.26	13.94
Other oxygen compounds[a]	0	7.22	5.29
Nitrogen compounds	traces	11.60	25.49
Sulfur compounds	0.45	2.40	0.20
Miscellaneous heterocyclics	traces	5.62	traces
Nonvolatile residue	0.00	22.80	25.80
Totals	100.00	100.00	100.00
Percent of atomic oxygen in hydroxy groups	0	30.3	80.2

[a]Includes tri- and tetraoxygen compounds as well as compounds containing both oxygen and sulfur atoms.

Table VII. Elemental Analysis of Asphaltenes Studied

Source of Asphaltenes	*Weight Percent*			
	Coal Extract	*Synthoil Product*	*EDS Distillate*	
	A[a]	A[a]	A[a]	B[b]
Carbon	85.85	86.82	82.13	83.52
Hydrogen	6.01	6.29	5.61	6.29
Oxygen	5.74	4.39	9.66	8.20
Sulfur	1.81	0.36	1.78	1.28
Nitrogen	0.97	2.11	0.79	0.71
basic	0.40	1.17	0.34	—
nonbasic	0.57	0.94	0.45	—
Total	100.38	99.97	99.97	100.00
Atomic H/C	0.835	0.864	0.820	0.904

[a]A: Conventional methods, including oxygen determination by neutron activation analysis.
[b]B: Calculated from MS composition of 77.20% volatiles.

Functional groups determined by IR techniques are shown in Table IX. The relatively small fraction of the total organic oxygen present in carboxylic and hydroxy groups indicates significant concentrations of etheric bridges and furanic functions. As one would predict from their thermal lability, acids and amides were found only in the asphaltenes from coal extracts.

Structural parameters calculated from NMR, GPC, and elemental analyses are summarized in Table X.

The prevalently aromatic character of these asphaltenes is confirmed by the 0.7 to 0.8 fraction of C atoms in aromatic rings. The ratio of aromatic to substitutable peripheral aromatic C atoms is about 1.50:1.60. This indicates an average condensation of four to five rings per unit structure, as in pyrene, $C_a/C_e = 1.60$, benzopyrene, $C_a/C_e = 1.67$, cholanthrene, $C_a/C_e = 1.50$.

The ratio of C atoms alpha to aromatic rings to peripheral aromatic carbon atoms is an indication of the degree of substitution. The approximately 30% value found for these asphaltenes corresponds to an average of three to four substituted aromatic carbons, if one assumes four to five ring aromatic structures that generally contain ten or twelve substitutable peripheral aromatic carbons (e.g., 10 in pyrene, 12 in benzopyrene).

Proton NMR cannot determine independently the number of naphthenic rings (R_n) and alkyl linkages (L), only their sum. Therefore, Table X contains three alternate possibilities for R_n and the related parameters, considering the changes in structural parameters corresponding to L = 0, 1, or 2. This is the most reasonable set of values for L, as it is unlikely that it exceeds the value of R_n and might be considerably less.

L values of 0, 1, and 2 correspond, respectively, to 1, 2, and 3 unit structures per molecule. This is consistent with limited UV data on the heavier asphaltenes that indicate one or two unit structures per molecule (*1*). UV data

Table VIII. Molecular Weight Distribution in Asphaltenes

Source	*Coal Extract*	*Synthoil Product*	*EDS Distillate*
Property			
Number-average MW	808	784	431[a]
Weight-average MW	1513	1483	741
Weight percent material in MW ranges			
<500	18	18	49
500–1000	24	27	32
1000–2000	37	37	15
<2000	21	18	4
Volatiles in MS inlet system	8	21	77

[a]Number average MW calculated from MS composition for 77% volatiles was 291. This indicates a value of 683 for the 23% nonvolatiles.

also exclude the presence of highly absorbing linear ring systems such as tetracene and pentacene. This implies more bulky condensed systems such as pyrene and benzopyrene, and this hypothesis is supported by the types of structures identified by UV in the cyclohexane-soluble fractions of coal liquids (*1*).

Table XI shows the comparison of MS and NMR data for the EDS distillate asphaltenes. The MS calculations are based on assigning a theoretical structure to each of the major individual carbon-number homologs determined, for example, assigning a benzopyrene type structure for the C_{20}–C_{29} carbon-number homologs in the C_nH_{2n-28} series. These assignments imply a certain number of rings and carbon atoms of the various types for each homolog. The $C_{22}H_{16}$ benzopyrene homolog is assigned, in this case, twenty aromatic C atoms, twelve peripheral C atoms, and two saturate C atoms, both in sidechains. Values assigned to individual components are then composited using their weight percent concentrations as weighing factors. Assumptions can be modified until a satisfactory agreement is reached with NMR values.

The composite values shown in Table XI were calculated assuming either fully aromatic or naphthenoaromatic structures with one naphthenic ring per molecule for the major homologs. Only one structural type was assumed for any given series, and this assumption could result in erroneously low values for naphthenoaromatics and high values for C atoms in sidechains. Nevertheless, the agreement obtained between MS and NMR is quite satisfactory.

The power of this combined NMR/MS approach is even more evident in calculating structural units and average number of sidechains.

The average MS hydrogen deficiency, 22.4, is compatible with the NMR ratio of about 1.50 for C_1/C_a only for L = 0. Although C_1/C_a as measured by

NMR is independent of the number of unit structures per molecule, the MS value for Z number is very sensitive to this parameter. L = 1, that is two unit structures per molecule, would be compatible only with an average Z value of 46, and this is definitively excluded by MS data on the volatile portion of the sample. We must conclude that L is equal to zero and assume 2.6 naphthenic rings per molecule (Table X). This conclusion could not be arrived at from MS or NMR alone, only the combination of the two techniques.

Combination of the NMR determination of 1.42 carbon atoms per sidechain with the MS value of 3.52 for average number of C atoms in sidechains yields a value of 2.5 for number of sidechains. This value is somewhat lower than that assumed from Table X for substituted aromatic C atoms (3 to 4), but it is not an unreasonable one.

We are now working on further extensions and refinements of the computer programs integrating NMR and MS data. One promising possibility is the elimination of the uncertainty by ^{13}C NMR between alpha to aromatic and junction carbon atoms by using values calculated by MS for the latter. Work in this area will be reported in further communications.

The data obtained on the asphaltene samples examined give considerable insight into both the average and the detailed structure of coal asphaltenes.

Table IX. Functional Groups in Asphaltenes as Determined by IR

	Weight Percent		
Source	*Coal Extract*	*Synthoil Product*	*EDS Distillate*
Oxygen			
in phenols	1.29	1.63	2.52
in hindered phenols	0.27	0.25	0.41
acid carbonyl			
free	0.08	Nil	Nil
associated	0.12	Nil	Nil
Acid OH			
free	0.08	Nil	Nil
associated	0.12	Nil	Nil
total by IR	2.27	1.98	2.93
total by NAA	5.74	4.39	9.66
percent accounted for	39.5	45.1	30.3
Nitrogen			
pyrrolic	0.19	0.40	0.09
amide	0.27	Nil	Nil
total by IR	0.46	0.40	0.09
total nonbasic	0.57	0.94	0.45
percent of nonbasic N accounted for	80.7	42.5	20.0

Table X. Average Structural Parameters Calculated for Asphaltenes from NMR, GPC, and Elemental Analysis

Source	*Weight Percent*		
	Coal Extract	*Synthoil Product*	*EDS Distillate*
Parameter			
fraction of aromatic C atoms	0.73	0.74	0.77
ratios			
aromatic C/peripheral C	1.59	1.48	1.48
C alpha to aromatic ring/ peripheral C	0.32	0.27	0.32
number of rings ($R_a + R_n$)	13.5	12.3	7.1
number of naphthenic rings (R_n) + alkyl links (L)	4.7	4.5	2.6
if L = 0			
R_a	8.8	7.8	4.5
R_n	4.7	4.5	2.6
unit structures	1	1	1
if L = 1			
R_a	9.8	8.8	5.5
R_n	3.7	3.5	1.6
unit structures	2	2	2
if L = 2			
R_a	10.8	9.8	6.5
R_n	2.7	2.5	0.6
unit structures	3	3	3
average number of C atoms per alkyl substituent	1.6	1.6	1.4

Table XI. Comparison of MS and NMR Data for EDS Distillate Asphaltenes

Parameter	*Determined by NMR on Total Fraction*	*Calculated from MS on Volatiles*	*Estimated from MS and NMR*
Average number of rings	4.5	>3.7	—
Fraction of aromatic C atoms	0.77	<0.82	—
Aromatic C/peripheral C	1.48	1.58	—
Average C number	—	20.15	—
Average number of C atoms in sidechains	—	3.52	—
Average number of C atoms per sidechain	1.42	—	—
Average hydrogen deficiency	—	22.42	—
Average number of sidechains	—	—	2.5
Average number of unit structures	—	—	1.0

The mystery of these materials is not resolved completely, and important questions such as hydrogen bonding vs. acid/base complexes remain to be answered, but at least we have some definitive understanding of the type of building blocks that are involved in the overall structure.

Acknowledgments

Work reported in this chapter was supported by the U. S. Department of Energy, Contracts No. E(46-1)-8007, E(49-18)-2369, EF-77-A-01-2893, the Electric Power Research Institute, Contracts RP-778-1 and RP-778-2, The Carter Oil Company (Exxon Local Company), Phillips Petroleum Company, Atlantic–Richfield Company, Japan Coal Liquefaction Development Company, Ruhrkohle A.G., and Exxon Research and Engineering Company.

Literature Cited

1. Aczel, T.; Williams, R. B.; Pancirov, R. J.; Karchmer, J. H. "Chemical Properties of Synthoil Products and Feeds," Report prepared for the US DOE, MERC 8007-1, 1977.
2. Aczel, T.; Williams, R. B.; Brown, R. A.; Pancirov, R. J. In "Analytical Methods for Coal and Coal Products"; Karr, C. Jr., Ed.; Academic: New York, 1978; Vol. 1, Chap. 17.
3. Chamberlain, N. F. "The Practice of NMR Spectroscopy"; Plenum: New York, 1974; pp. 56–66.
4. Williams, R. B. *ASTM Special Technical Publication,* **1958** *224,* 168.
5. Williams, R. B.; Chamberlain, N. F. *World Pet. Cong., Proc., 6th, 1964.*
6. Van Krevelen, D. W.; Schuyler, J. "Coal Science"; Elsevier: Amsterdam, 1957; Chap. 7.
7. McKay, J. F.; Cogswell, T. E.; Weber, J. H.; Latham, D. R. *Fuel* **1975,** *54,* 50.
8. Snyder, L. R. *Anal. Chem.* **1969,** *41,* 314.
9. Petersen, J. C.; Barbour, R. V.; Dorrence, S. M.; Barbour, F. A.; Helm, R. V. *Anal. Chem.* **1971** *43,* 1491.
10. Scheppele, S. E. et al. *Proc. Ann. Conf. Am. Soc. Mass Spectrometry, 26th,* St. Louis, MO, 1978, 64–66.

RECEIVED June 23, 1980.

INDEX

INDEX

C

D

E

Jacket design by Carol Conway.
Editing and production by Robin Giroux and Cynthia E. Hale.

Composed by Science Press, Ephrata, PA, and Trade Typographers, Inc.
Washington, DC
Printed and bound by R. R. Donnelley & Sons Co., Chicago, IL.